TRENDS IN COMMUNICATION

VOLUME 12 | NUMBERS 2&3 | 2004

Sport and the Media: Recent Economic, Legal, and Technological Developments

Raymond Boyle, Peter Flood, and Deirdre Kevin
Issue Editors

CONTENTS

First published 2004 by Lawrence Erlbaum Associates, Inc.

Published 2019 by Routledge
2 Park Square, Milton Park, Abingdon, Oxon OX14 4RN
52 Vanderbilt Avenue, New York, NY 10017

Routledge is an imprint of the Taylor & Francis Group, an informa business

This journal is abstracted or indexed in *EBSCOhost Products* and *Communication Abstracts.*

ISBN 13: 978-0-8058-9522-3 (pbk)
ISSN 1383-8857.

TRENDS IN COMMUNICATION *12(2)*, 69–71 (2004)

Editorial

Raymond Boyle
Film and Media Department
Stirling University

Peter Flood
Media Consultant
Düsseldorf, Germany

Deirdre Kevin
European Institute for the Media
Düsseldorf, Germany

When media mogul Rupert Murdoch identified sport as the battering ram for pay-TV in the mid 1990s, he may well have had football specifically in mind. For although Murdoch is not thought to be particularly interested in sport, he has always from his early newspaper days recognized its key commercial value as a driver of sales in the media marketplace.

For most of the major European media markets, football above all others sports has been the key "media content" over the last decade or so, as these markets have been systematically opened up to competition through a range of technological, economic, and political factors.

Murdoch's BSkyB pay-television station in the United Kingdom is one of the most profitable in the world, with its 7.2m Sky Sports subscribers driving its core revenues. Since 1992, the company has paid almost £3billion to secure the exclusive live rights to top flight English football, and its coverage of the game has become synonymous in the United Kingdom with the BSkyB brand. Despite the downturn across Europe of the value of sports rights since the dotcom highpoint in 2000, football in particular remains at the center of the content packages being rolled out across a range of digital media platform—digital television, the internet, and mobile telephony—as companies compete to secure subscribers and advertisers.

The growing adoption of personal video recording (PSV) systems, a side effect of digitalization, is only the latest in a succession of technologically led threats to the advertiser's ability to address a desired audience. Sportscasting, and to a lesser extent newscasting, remain the most reliable means of assembling a mass audience. From the first wave of proliferation driven by cable and satellite broadcasters, sport has been established, or at least widely regarded, as the "killer application" to

Requests for reprints should be sent to Deirdre Kevin, European Institute for the Media, Zollhof 2a, Düsseldorf D-40221, Germany. E-mail: kevin@eim.org roll-out and sustain successive media, outlets, and business models. Now, 4 years after the zenith of the dot com bubble, and two years after the collapse of Kirsch media and ITV digital, multinational media, and telecommunications. Enterprises seem to be steadying themselves for the next period of growth. It is against this background that *Trends in Communication* presents this issue on "Sport and the Media: Recent Economic, Legal, and Technological Developments." The articles examine just how reliable sportscasting is in this anchor role, and review the development of sportscasting and specialized sporting services as a response to audience demand, and question how well the sporting enthusiast—the fan—is being served by later, unanticipated, developments. For although sport has had a long symbiotic relationship with communication systems (Boyle & Haynes, 2000; Rowe, 1999), the advent of a digital media age has intensified that relationship giving us evidence of both continuity and also substantial change in this relationship as both media systems and popular culture have entered a new century.

The first article from Solberg and Hammervold presents a salutary reminder of the financial pain suffered by broadcasters in many territories by letting an overexcited market determine the value of various sports rights. The article goes on to examine and analyze the mechanics of what it is that make sports rights in particular so compelling to broadcasters: the declining phenomenon of the mass audience. Looking at this phenomenon, the authors isolated characteristics of individual football matches within a tournament and the impact these have, both in attracting an audience, and in sustaining their interest in commercial messages embedded in the coverage. This work represents a new analytical and scientific approach towards managing investment in sports rights in order to maximize profit.

Raymond Boyle reviews the history and prospective future of sports coverage as a leading content application in the rollout of the 3G mobile telecommunications. No less than Solberg and Hammervold, Boyle questions whether sport will, in the final analysis, be central in the rollout of this costly technology. This article places a realistic perspective on any economic expectations of this technology and indeed related expectations for exploiting sports content. Those expectations hinge upon exploiting a new technologically defined sports-content product in the form of 3G. dedicated coverage. The commentary additionally places the potential impact of 3G technology within some of the broader discussion here regarding sale of sports rights and regulation of competition issues in terms of access to, and bundling of these rights.

This process of subdividing rights within sport is the subject of Haynes article that explores the recent history, and legal nature, of *image rights* in sport. Such rights, more familiar in the United States, are known to some extent, and recognized in the legislation of various states of the European Union (EU). Haynes illustrates how without legislation but through custom, contract, and regulatory forbearance these rights are becoming recognized in the United Kingdom. The emergence of image rights is, again, symptomatic of the huge sums of money that have been invested in British soccer since the beginning of the 1990s. This investment has had a very substantial impact on the way the game is experienced by its enthusiastic supporters.

Dahlén and Thomsen, trace the development of sports coverage of radio and dedicated sports channels from the mid-1980s. Using the example of Norwegian radio, they describe the impact of gathering commercialization on the quality of sports coverage. The article outlines two opposed format or business models that have been applied to sports broadcasting by local radio networks when streamlining services where different use is made of the potential revenue from sport. In so doing, they chart at various points the denial or removal of services from an established audience.

This matter of commercialization denying service to audiences is at the heart of Roderick Flynn's article that completes this issue. In 2002, and within weeks of the completion of the World Cup finals, it was announced by the Irish Football Association (IFA) that exclusive rights to home games of the Irish national soccer side had been sold to Rupert Murdoch's BSkyB. Although it was initially unclear whether these games would be shown on a premium channel received by only 15% of the Irish audience. This purchase, and the apparent impotence of the national broadcaster RTE, caused something of a sensation in Ireland at the time. In particular the Irish government moved both to retroactively "list" the games in question and to move the regulation of transnational broadcasting onto the EU agenda in the area of transfrontier television. Additionally it elucidates the relationships between political and media actors, and the cultural and social backlash from the sale of rights that forced the politicians to compromise this important relationship.

These articles collectively give an authentic flavor of the issues, real and incipient, surrounding sports broadcasting, the media, and the audience as the next round of media consolidation and investment appears to be starting. In particular they pose questions as to whether sports as a "killer application" has actually encountered some natural limit, and whether the audience at-large has been well served within the more recent developments.

REFERENCES

Boyle, & Haynes. (2000).
Rowe. (1999).

TRENDS IN COMMUNICATION *12(2)*, 73–82 (2004)

Mobile Communication and the Sports Industry: The Case of 3G

Raymond Boyle
Stirling Media Research Institute
Stirling University

This article focuses on mobile communication and sports content and examines what the experiences of the interplay between sport and new media tell us about the emerging and developing relationship between sports content and mobile telephony. The research looks at the role that sports content has played in the evolution of third generation (3G) mobile handsets and questions its viability as "killer content."

I argue that mobile media has long been important to the sports fan, specifically those who attend a live event and that while much will change over the coming years as mobile technologies become more deeply rooted in our everyday lives, there will also be substantial aspects of continuity in how sports fans use media.

3G, telecommunications, new media, sports, football

Major global events such as the Olympic Games are highly prized by content providers and can be a driving force behind the roll-out of new communication services, be it through the Internet or via mobile telephone networks.

—European Commission Chief Mario Monti, February 19, 2004

1. INTRODUCTION

This article focuses specifically on mobile communication and sports content and examines what lessons we can learn from the relationship that has evolved between sports and various new media platforms. This particular research is primarily concerned with the mobile telecommunications sector of the new media economy, while recognizing that the various platforms across digital media are interconnected. Indeed this is one of the supposed key characteristics of this particular stage in the evolution of the media industries.

Debates about the role that sports content plays in the new media age, specifically around platforms such as the Internet and digital television, have been covered in some detail elsewhere (Boyle & Haynes, 2004). Although they clearly impact on the specific area examined in this research, the particular focus of attention here is on what the experiences of the interplay between sport and new media

Requests for reprints should be sent to Raymond Boyle, Stirling Media Research Institute, Stirling University, Stirling, FK9 4LA, Scotland. E-mail raymond.boyle@stir.ac.uk in these sectors of the digital economy tell us about the emerging and developing relationship between sports content and mobile telephony and 3G networks specifically. Much of the discussion will focus on football, as it has been identified as the "key killer concept," and the most important sports content across a range of media platforms. This process has been assisted with the identification of the key demographics of the football fan as being mostly male, young, and with discretionary income: a classic profile of the early adopter of new mobile technologies. Although the wide spread appeal of the sport beyond this group is important for various media outlets, this core group remains key to new media providers.

There is of course a long history between sport and new technologies. It was sport that was one of the key content providers in the early days of both radio and television. Other televisual technological innovations are also associated with sports content, the first transatlantic live satellite link saw Europe view baseball from Chicago, the BBC television coverage of Wimbledon in the late 1960s saw the introduction of color, whereas the video tape was first introduced into the United Kingdom by the BBC in 1966 for the FIFA World Cup, thus facilitating the first use of slow motion action replays.

However the new media digital age is increasingly one driven not by issues of public service, but rather by commercial values and a wider economic climate in which the market remains the central driver of the digital economy. This factor will be addressed in the concluding part of this article.

In early 2004 the ongoing global economic downturn is still adversely impacting on media/advertising budgets in the United Kingdom. Media spend, which is the lifeblood of commercial television, fell in 2003 by 2.6% in a sector already suffering from a number of years of declining revenue in real terms. The growth of broadband usage in the United Kingdom is also rising, passing the 3 million mark at the end of 2003. However the use of this term is confusing as connections range from 160k to 1mb. Put simply, many such packages are not good enough to watch live-streamed sport. Additionally, in what could be described as a case of back to the future, November 2003 saw AOL UK strike a deal to show football on the Web with PTV (40,000 subscribers) placing football on the Web at the heart of AOL UK's broadband strategy. Some of us remain skeptical of such a strategy, and indeed whether the broadband market will continue to grow at its current pace is difficult to predict, as the numbers wanting to upgrade from dial up, reach saturation in the United Kingdom. It is against this developing backdrop that the mobile sports industry finds itself operating.

2. THE SPORTS NEW MEDIA NETWORK 1999 TO 2004

A series of snapshots from the media sport landscape of the last few years throw out some interesting images. It is 1999 and the height of the new media gold rush and a feeding frenzy exists between sport and media platforms that view the former as key content. Both old and new media companies invest millions in football clubs as they attempt to grow and create businesses through sports content. It is envisaged that a new media sports landscape characterized by converging technologies and a level of interactivity previously unimaginable is emerging that will restructure our sporting experience for ever.

In the era of "content as king," football clubs were busy building up their media capacities and waking up to the value of image content that they had up until this point never fully exploited. Clubs were hoping to use the new technologies of the Internet, digital television, and mobile telephony to commercialize all areas of their activity (Boyle & Haynes, 2004). Sport, and football in particular, was viewed as the "killer content" required to drive subscriptions and uptake across a range of media platforms.

Then the dot.com crash and burn (Cassidy, 2002; Cellan-Jones, 2001), allied with a spectacular failure of digital television to secure an adequate return on the massive investment in football rights. The collapse of ITV Digital in the United Kingdom and Kirsch Media in Germany, being two high profile examples of this failed strategy, with the demise of both companies inflicting extensive collateral damage on the sport in both England and Germany (Boyle & Haynes, 2004).

It is 2004 and those companies still left standing are systematically restructuring their relationship with sport. Rumors abound in the industry that Granada Media are trying to extricate themselves from their deals with Arsenal and Liverpool football clubs. The telecom company NTL Europe, set up when NTL went technically bankrupt, takes a £23m hit to sever its links with just one football club, Aston Villa. Even that most canny of companies BSkyB right off £300m it paid for Web operator Sports Internet.

When one examines the digital sports market in 2004, a number of trends emerge. In the general digital economy there have been clear signs that the slump in new media telecom stocks may be coming to an end, with evidence suggesting that once again this sector of the economy both in the United States and Europe is about to embark on a period of growth (Walsh, 2004). Significantly there have also been indications that the online media market is reaching a stage of maturity with a number of increasing company mergers as a process of consolidation takes place.

Interestingly this suggests that despite initial predictions, many of the basic business tenets applicable in the offline market are being reproduced in the online media economy. In other words, up to five major businesses are tending to dominate specific online markets in much the same way as other sectors of the economy. As Seb Bishop cofounder of Espotting argued, "I think we are now starting to see very old school values applied to a new medium" (Hargrave, 2004, p. 36). Of course, crucial in mapping and shaping future developments will be the wider economic and political context in which the evolving communications industry operates. In the United Kingdom, for example, despite a slow up take, digital television has been boosted with the launch in 2003 of the BBC led free to air digital terrestrial system, Freeview, which was in 2.5 million homes by December 2003. With satellite and cable digital systems already in up to 10 million homes, the U.K. digital market is beginning to bed down.

Clearly the creation of a mass digital television market place (with both a robust free-to-air and pay sector) will dramatically impact on the way in which sport is mediated, bought, sold, and packaged. While the battle to gain a significant foothold in the evolving mobile market continues.

3. THE MOBILE MARKETPLACE: FROM 2.5 TO 3G

The relative saturation of the mobile phone market in many of the key European markets is helping to drive the continual desire of the main telecoms players to create and establish new and more lucrative revenue streams with the roll out of new technologies and attendant services. It is estimated that mobile penetration levels are reaching 80% in European markets (Pentland, 2003) and the development of the next generation of mobile phones from 2.5 to 3G are seen as an important aspect of the long term growth of telephony revenues across Europe.

The allowing of data streaming and picture messaging through 2.5G (second generation enhanced phones) has clearly extended the range of services available via mobile technologies. The launch of the 2.5G service Vodafone Live, by the largest telecoms company in the world in October 2002, heavily promoted by football icon David Beckham, saw the company sign 3 million customers in the first year. It also allowed the company to bed down the technology and test the marketplace.

However the next generation mobile revolution will, it appears be around 3G or UTMS technology that allows the streaming of all the above, plus video, and the facility to video conference. 3G can deliver mobile video content at between 84 to 128kb a second, as opposed to around 46Kb on 2.5G phones. This allows, for example, higher quality clips of between 2 to 2.5 minutes of football match highlights to be delivered across a 3G network.

The first 3G mobile phone was launched in Japan by that country's leading mobile operator NTT DoCoMo in October 2001. However despite the franchises being auctioned by the U.K. Treasury for a massive £22.5bn early in the new century, in Britain it would be March 2003 before the much delayed launch of 3, the first 3G network in that country (and one of the first in Europe) took place. Originally the company Hutchison 3G (co-owned by Hutchison Whampoa, KPN Mobile from Holland and NTT DoCoMo in Japan) aimed to market their range of services, with sports content including exclusive football highlight packages featuring the English Premier League, at the core of its customer appeal.

The actual deal with the Premier League ran from the start of season 2001/2002 at a cost of around £45m, however technology problems around the launch of the network meant that value could only start to be extracted from these rights at the tail end of the 2002/2003 season. Following slow up take of the services in 2003, 3 began to reposition their marketing around cost and the inclusion of free content in order to drive uptake. They also launched a pay as you go service (something originally discarded in their business plan). However as research by the Work Foundation in the United Kingdom (Crabtree, Nathan, & Roberts, 2003) clearly demonstrated, the cost of the new range of mobile services remains an important factor and concern for many mobile consumers.

By March of 2004, 3 had 361,000 subscribers in the United Kingdom (they originally hoped for 1 million by December 2003), with Hutchison's Italian service having 453,000 customers (Gibson, 2004). With the entry into the fray across Europe in 2004 of 3G networks from Orange, Deutsche Telekom's T–Mobile, and Vodafone, competition in the market is set to become more intense and the need to attract and retain customers to the networks more difficult. This is brought into sharp relief when you note that across the continent telecom companies have paid over £50bn for 3G licenses. For many involved in the business, sports content remains a vital part of the services mix needed to entice customers to this market.

4. SPORTS CONTENT AND MOBILE COMMUNICATION

There is however growing evidence to suggest that despite the range of deals that have been struck between companies and sports related content providers that the value of sports mobile content may have been overstated. Tahoe Networks (2004) research indicated that only 18% would pay to have personalized sports content on their 3G phone and the figure only rises to 33% of those who wish to have sports updates sent to their mobile or PDA.

However like much of the wider new media environment, the sports mobile media content market has been characterized by consultancy reports keen to identify the riches that lie in store for sports clubs and media operators when sports content and new media synergize. Sports have been identified as a core strategic "content" that is being used to attract subscribers to the new 3G/UMTS services as they begin to roll out across Europe. This is a vital stage in the development of this sector of the new media economy. The UMTS Forum (2003) optimistically estimated that this market will be worth $164bn by 2010, with the average European subscriber spending up to 170€ a month on such services.

Although the significant technological investment in developing a robust platform has been important, without the delivery of a range of key services, which enhance the growth of the telecommunications market then sustained long term competitive growth in this media sector will be difficult.

However, more recently some consultants and media analysts have been engaged in a retrenchment of sorts. Lars Becker, Chief Executive of mobile marketing agency Flytxt, for example, argued that sport is not that important, and he expected 3G killer content is more likely to be music, adult services, and gambling (Wood, 2004).

Indeed one of the core problems faced by 3G operators has been identifying what is the key content that will drive subscribers and adopters to the service. There is no particular agreement across the emerging 3G European market as to the correct mix of services that will entice customers to invest in the new services.

Although sports content has proven successful in driving uptake in various pay television markets across Europe (in particular the U.K. and French markets), its success has been less marked in the new digital television markets (Boyle & Haynes, 2004). However simply transporting this model across media platforms to the Internet (either dial up or broadband) and those mobile telephony markets has been less successful. A failure to understand how the sports fan engages with these technologies and how the architecture of new media are not necessarily suited to the wider cultural and social patterns of sports consumption has led to mismatched expectations for both media companies and sports clubs who both envisaged a combination of sports content and digital technology driving new revenue streams towards them.

5. REGULATING THE MARKET

It is also clear that the European Commission's (EC) interest in the relationship between new media, sport, and sports rights will intensify over the next decade, particularly as they seek to create a climate in which the new media economy can flourish. The start of 2004 saw the Competition Commission launch a major investigation into the relationship between sports content and the 3G mobile phone networks across Europe. At the heart of this study is a belief that the success of these commercial networks requires open competition for key content of which sports is earmarked as one such area.

There are already wider implications, for public service television sport, of the EC's 2003/2004 attempts to restructure deals struck by the European Broadcasting Union (EBU), which buys events such as the Olympics and FIFA World Cup for free-to-air public service broadcasters throughout Europe. The EC are challenging the EBU's right to buy sports rights collectively claiming it is anticompetitive and part of a cartel that excludes other broadcasters. However, should the EC follow through on this, then the door opens for commercial broadcasters to simply cherry pick top events, again undermining public service television.

Despite cultural differences across various European audiovisual markets, football remains one of the most universally important sports in such media markets. This has been recognized by the EC as it continues to investigate the selling of television rights across European markets and in particular the role the sport plays as key content on digital television platforms. Thus there are already significant moves by telecommunications companies to secure premium sports (and football in particular) content agreements.

Significantly, the area of sports rights in the digital age is itself a complex arena with a range of key stakeholders who are seeking to maximize the value of the image rights they feel constitute part of their intellectual property portfolio.

Thus the issue of 3G rights needs to be alert to the complex debates between football clubs, national football associations, and pan European organizations such as UEFA who preside over the European game. Many elite clubs such as Manchester United, Real Madrid, and Juventus are keen to retain and exploit as many new media rights across a range of digital platforms such as 3G/UTMS, whereas national and European governing bodies are keen to retain some element of collective selling. Thus the bundling and unbundling of 3G/UTMS rights is of central concern to a range of stakeholders across both European sporting and new media markets, with a number of elite European football clubs already entering into 3G content arrangements.

The development of secondary and tertiary markets for sporting content on UMTS and broadband networks across Europe, although uneven, has frequently been characterized by concerns and questions related to market dominance, bargaining power, bundling or conditional selling, and other issues related to competition. It is always difficult to predicate how such patterns of regulation will develop in practice. However it is clear that the future evolution of 3G sports content will be as much about evolving battle lines between elite football clubs, as "upstream" media content providers and media organizations as "down stream" distribution systems, as both parties attempt to reposition themselves as both providers and deliverers of football content.

6. THE STATE OF PLAY: EVOLUTION NOT REVOLUTION

The history of communication innovation is one littered with technological developments that roll out on a wave of hyperbole. A period of market testing and often retrenchment follow, and some technologies become embedded in our everyday lives for a range of social, cultural, and economic reasons.

Mobile media have long been important to the sports fan, specifically those who attend a live event. The radio has been a significant vehicle for supporters at football matches as they seek immediate information about games taking place elsewhere. Indeed the general decline in supporters carrying radios into football grounds indicates that for many the function of keeping in touch with other games has been taken over by the mobile phone (with of course recent models now incorporating a radio into their range of functions). SMS is also an increasingly important part of the live sporting experience, not only for receiving information about results etc, but also as an organizational aid as supporters arrange to meet socially before and after matches.

As far back as the European Football Championships in 1992, English ticket touts were using mobile phones as a key business tool as they plied their illegal trade around the stadia of Sweden during that international sporting event. What is central here however is the way that technology has added to, or enhanced the experience for the supporter. In other words, there has been a dovetailing of the needs of the fan with the technological milieu available to them.

Given the importance for any successful business model that meshes the social practice of the supporter with the range of services offered by mobile handsets, it seems that there are limitations associated with the current visions being articulated about how fans will use 3G technology to enhance their sporting experience. One issue is that of exclusivity. Will fans want to watch, and more important pay for, goal clips and short highlights of games that they will be able to watch on a television screen later that evening? While Euro 2004 in Portugal was the first major football event to allow immediate post match delivery of such services (up until then fans with 3G had to wait until up to an hour after the game), the argument still stands. Thus the issue of exclusivity is set to become more important within the mobile market, as it has done in the digital pay-TV sector over the last decade or so. However, given the relatively small size of the current 3G market, it is clear that rights sharing between networks will remain for some time to come.

For fans at the game, most major clubs will now anyway show highlights and the goals on large screens in the stadia after the game. Given the plethora of radio sports phone-in programs will fans pay to see postmatch interviews that will often be carried on local radio programs?

Mobile phones have become an increasingly important part of the wider social practice of attending live sport, as an organizing aid, as a source of information and even as a way of connecting with other friends not at the live event (phones being used to relay the atmosphere at major football matches is not an uncommon sight at big games). Postmatch they allow easy access to the interactive radio phone-in programs that review the incidents and action from the events. However, the range of services currently being offered through 3G technology, in particular live streaming of visual sports material, does not really mesh with the needs of supporters. At present we appear to have a technologically driven view of what 3G can do, which fails to understand or appreciate the importance of the distinctive social environments within which the vast majority of sports fans find themselves, while recognizing that this can significantly differ from sport to sport.

Another, often underrated factor in the uptake of 3G technology is the issue of cost. Research by the Work Foundation in the United Kingdom (Crabtree, Nathan, & Roberts, 2003) identified that the cost of mobile services, specifically in the domestic market, was a key factor that inhibited upgrading and the take up of new mobile services. In other words, any new service must be both compelling and enhance already existing social patterns of new media consumption. The cost of being fully connected to the new media economy is not cheap. Although broadband costs are falling in the United Kingdom, a connection, when added to a digital pay-TV package and a contract mobile account means one can find themselves paying on average anything between £35 to £70 a month—a considerable impact on domestic budgets.

This leads to another related aspect of the new media sports ecology, this is the extent to which it has been driven by an overwhelmingly commercial consumer model of development. From the interventionist concerns of the EC discussed above, through to the sports content holders and the telecom providers, the key challenge has been to identify services that will make money from these new technologies, rather than an interest in the potential beneficial social and cultural impact that digitized media may have on our lives.

The absence of the public sector from the sports new media debate has been striking. Where public service broadcasters such as the BBC have been involved, through their hugely successful BBCi sports Website in the United Kingdom, they have been attacked by commercial rivals for their distortion of the new media market. Yet what they have been involved in is extending and reformulating the definition and scope of public service in a new media digital age. This remains a considerable challenge and is part of a wider communicative process that Tambini and Cowling (2004) argued is required for the "renewal of public service in the digital age" (p. 4).

Ironically, the failure of the Internet to deliver the promised riches to the football industry in the United Kingdom (Boyle & Haynes, 2004) has lead some football clubs and content providers to take a more innovative strategic approach when examining what the Internet can do for their club. Thus a return to recognizing the inherent strengths of the Internet and its ability to help consolidate and build community relations among their key fan base has been the driver for clubs such as Charlton FC in the English Premiership, rather than it simply being a vehicle to deliver additional revenue streams to the business (www.charlton-athletic.co.uk).

7. CONCLUSIONS: EVERYTHING TO PLAY FOR?

We are clearly at the start of the next wave of new media developments as they interface with sports and sports culture. Companies such as Vodafone, who are highly immersed in the sporting world from Formula 1 motor racing, through horse racing to football have a 20 year business plan to drive returns from the 3G license they secured in the United Kingdom for £6bn. Although they see 2005 as an important landmark in growing 3G business, they are in this sector for the long haul (Cronin, 2004).

Yet identifying future trends of new media consumption has already proven to be inherently difficult to predict, as the way these digital technologies become embedded in our everyday social lives has not always been in a manner that has been easily envisaged. The rise of SMS on mobile phones in a supposedly overwhelmingly visual age, is one such example. On November 22, 2003, 76 million text messages were sent in the United Kingdom, the third highest number of the year. It was the day England won the Rugby World Cup and SMS has proven to be one of the areas in which telecoms are generating significant revenues.

For example, 2004 was being predicted by the new media consultants (the same ones who got it so wrong in 1999) as the year of wireless Internet connections (Wi–Fi), offering public locations where people can simply open their laptop and connect. The potential structural impact for existing payment structures and competition of such developments that would eliminate local fixed lines are potentially enormous. Yet how realistic a mass market scenario Wi–Fi actually is, remains open to debate.

In January 2004, British Telecom entered a partnership in the United Kingdom with the fast food chain McDonalds to allow Wi–Fi access in many of its outlets creating the largest wireless network in the United Kingdom, yet is one likely to simply take out an expensive laptop in a McDonalds located in a crowded city center and be able to concentrate without fear of someone coming along and stealing it? Why would one want to spend time in a fast food restaurant environment that is designed to encourage you to eat quickly and leave, surfing the Internet and answering emails? Given that these are commercial arrangements, what exactly are the revenue benefits for such companies? This process of trial and error is likely to continue as we find our way around the digital landscape and adapt it to suit our own lifestyles and patterns of popular cultural consumption.

Consultants such as Spectrum remain convinced that sports services will be a key driver for mobile growth over the coming years (www.spectrumstrategy.com), at the core of their analysis is the need for "the creation of 'a rich multi-platform experience (across TV, internet and mobile)'"(Spectrum, 2002). Yet even in 2004, the wider architectural shape of any converged new media edifice remains unclear. Although convergence is happening, it has not always been in the ways envisaged, and it occurs at a much more uneven pace and pattern. For sports fans, television, either in the domestic or public environment, will remain the key technology for delivering sport outside of the live arena for the immediate future. The Internet and mobile communications offer additional enhancements to the sporting experience and in some cases, such as in the provision of information, they have significantly altered the media sports relationship in fields such as journalism (Boyle & Haynes, 2004).

Clearly much will change over the coming years as mobile technologies become more deeply rooted in our everyday lives, however there will also be substantial aspects of continuity in how sports fans use media. The previous waves of media innovation from the print and broadcast media suggest a sense of a continuing evolution in the relationship between technologies, sport and supporters rather than the full scale revolution that was so widely predicted.

REFERENCES

Boyle, R., & Haynes, R. (2004). *Football in the new media age.* London: Routledge.

Cassidy. J. (2002). *dot.con: The greatest story ever sold.* London: Allen Lane.

Cellan-Jones, R. (2001). *dot.bomb: The rise and fall of dot.com Britain.* London: Aurum.

Gibson, O. (2004, March 18). 3G sales rocket. *The Guardian.*

Crabtree, J., Nathan, M., & Roberts, S. (2003). *MobileUK: Mobile phones and everyday life.* London: Work Foundation.

Cronin, J. (2004). 3G—A mobile revolution? Retrieved from news.bbc.co.uk/2/hi/business/3590055.stm

Hargrave, S. (2004, February 16). On the right track. New Media. Media Guardian. *The Guardian.*

Spectrum Strategy Consultants. (2002.). *Translating TV sports into mobile* (Tech. Report). Retrieved from www.spectrumstrategy.com

Tahoe Networks. (2004). Retrieved from http://www.tahoenetworks.com

Tambini, D., & Cowling, J. (2004). *From public service broadcasting to public service communications.* London: IPPR.

UTMS Forum. (2003). *Mobile evolution: Shaping the future.* London: Author.

Visiongain. (2003). *Increasing revenue through wireless sport content.* London: Author.

Walsh, C. (2004, February 8). Fit to pop? *The Observer.*

Wood, D. (2004, March). Killer content on 3G, where content meets mobile. *Broadcast Magazine.*

TRENDS IN COMMUNICATION 12(2), 83–100 (2004)

Sport Broadcasting: How to Maximize the Rating Figures

Harry Arne Solberg
Randi Hammervold
Trondheim Business School
Sør-Trøndelag University College

This article analyzed how the *profile* of the teams and *quality* of the matches influence the rating figures during the broadcasting of live soccer matches. It confirmed that national teams (Rosenborg in this case) are especially popular, but also that other teams can be extremely popular (Manchester United and Liverpool in this case)—independent of the viewers geographical origin. When such teams are eliminated from a tournament—or do not even qualify—this can hit the TV channels financially. The data also revealed that even if soccer matches attract high rating figures, this is no guarantee that the prematch sections will also attract high rating figures. The median viewer of Champions League matches started to watch when the match was about to start, and thus did not take any interest in the prematch commercials. Hence, the optimal solution for profit maximizing channels may be to start the program shortly before the match. If there are few who watch the long prematch and postmatch sections, then the alternative costs of screening them will be high, particularly if the channel could have alternatively broadcast other attractive programs.

revenue maximization, regression analysis program scheduling, sports rights, audience preferences, ratings analysis, soccer matches, profile of teams, quality of matches

1. INTRODUCTION

In Europe, the markets for sports rights have gone through several periods of turbulence since the early 1990s. Since then the prices of the most attractive products have increased substantially. Simultaneously, a number of rights' deals have turned out to be unprofitable for the acquiring channels. The bankruptcy of ITV–Digital and Kirch Media are well-known examples. Several TV channels suffered losses from broadcasting the 2002 World Cup soccer finals. The Italian channel, RAI, threatened FIFA with legal proceedings when Italy was eliminated from the tournament because TV pictures revealed that the referee mistakenly annulled a goal that would

Requests for reprints should be sent to Harry Arne Solberg, Trondheim Business School, Sør-Trøndelag University College, Jonsvannsvegen 82–N–7004 Trondheim, Norway. E-mail: Harry.solberg@toh.hist.no have kept Italy in the tournament. Furthermore, the French channel, TF1, lost €18 million due to the early exit of the French team (Desbordes 2003). Canal Digital, a Nordic channel, lost NOK48 million from broadcasting the tournament (http:/ www .telenor.com/ir/quarterly_reports/2q02/pdf_xls/2_kvartal_2002eng_godkj.pdf). Similar problems have hit European channels that have broadcast matches from UEFA Champions League when national clubs have been eliminated at early stages—or not even qualified—and hence reduced the rating figures in the respective markets.

Such problems have occurred regularly in North America where sports rights have a longer history of being a commodity than in Europe (Solberg 2003). Several networks suffered severe losses due to the September 11, 2001 aftermath, as it reduced advertising on TV dramatically. In 2002, News Corporation, a media company that owns several television networks (i.e. BSkyB and Fox) took a write-off of $909 million on unprofitable sports rights deals, whereas NBC (one of the major broadcasting networks) reported losses on their National Basketball Association deal (NBA) estimated at $300 million over two seasons (http://www.gouldmedia.com/tsr.html). In the beginning of the 1990s, CBS, another network, almost went bankrupt when they bid too much for the Major League Baseball (MLB) rights (Fort, 2003).

These examples illustrate that TV networks, which invest in sports rights, are running a high risk of being hit by the *winner's curse*. This refers to a situation in which the winner of a common-value auction is worse off as a consequence of overestimating the value of the item and thereby overbidding (Pindyck & Rubinfeld, 2001).

In this article we estimate which variables (and how much) such as the *profile* of the teams and the (expected) *quality* on the matches influence the size of the TV audience at live sport programs. Important research questions are:

- Which categories of teams and matches attract the highest numbers of viewers?
- What is the relationship between the rating figures during the matches and during the commercials?
- What is the optimal time to start and end the programs?

Sports rights are often sold as bundles, which include a large number of products, for example, the matches in a tournament. This mixture of products—combined with the fact that many contracts run over several years—make it difficult to estimate the rating figures at the point of time when the contract is settled. The interest for the specific products can vary considerably and can be influenced by factors such as:

- *Specific teams/competitors:* Many viewers are mainly interested in watching their own national favorites and loose the interest for the tournament once these favorites are eliminated. Their interest can be stimulated if these participants perform well prior to the tournament or in its first stage. Furthermore, some teams have a special popular status among many viewers—which is independent of their geographical origin—for example, the Brazilian national soccer team.

- *Alternative goods (substitutes):* The number of TV channels has increased considerably in recent years, partly due to the new digital technology. Hence, the viewers are offered more alternatives than ever before. This includes sports programs as well as other entertainment programs. In addition, TV also has competition from other leisure activities.
- *The point of time (period) when the programs are broadcast:* It can be difficult to achieve high rating figures if the tournament is hosted on a continent with a different time zone. Furthermore, a collision with the season of other attractive sports can also reduce the numbers of viewers.

The empirical section presents regression analysis where the dependent variable is the rating figures during soccer matches as well as during the commercials related to these matches. Before that, however, the next section presents some characteristics of broadcasting in general, and clarifies some theoretical principles regarding the construction of a menu of programs. Special attention is given to how to maximize the rating figures—the income side. This is followed by some empirical analyses based on data from the broadcasting of live matches from UEFA Champions League in Norway. We will use a multiple regression model to estimate how specific variables influence the rating figures during the commercials and the matches. The final section discusses how commercial advertising channels can use this information to maximize the advertising revenues from broadcasting sport programs.

2. THEORETICAL CONTEXT

2.1 TV Broadcasting—Some Characteristics

With its high degree of fixed costs and relatively low degree of variable costs, TV broadcasting is a typical *economies of scale* production (Solberg 2002). The production and transmission of TV programs requires considerable initial funding. Once the initial costs have been recovered, those that can afford the entrance costs will enjoy economies of scale and scope by spreading overhead costs, applying common managerial techniques, spreading risks, and enjoying pecuniary advantages such as discounts from program suppliers (Dunnett 1990).

The broadcasting of TV programs also contains some elements of *public goods*—both in the production and in the transmission process. The reception of a television signal by one person within a broadcasting market does not prevent it from being received by another. Once the programs are being transmitted, the costs are independent of the number of viewers, and can be regarded as sunk costs, irrespective of whether they are watched by a thousand or by a million viewers. To the broadcaster, the marginal cost of transmitting a program to one additional viewer within any broadcasting market area is literally zero (Gaustad 2000). This corresponds to the nonrivaling criterion characteristic of public goods (Olsson 1971). Hence a channel cannot reduce the costs by broadcasting the programs to fewer viewers in case the demand from its advertisers is hit by a negative shift.

These characteristics make the channels extra vulnerable to negative shifts in demand from viewers and advertisers. To avoid—or at least reduce the risk of being hit by the winners' curse—the channels need information about the factors that influence people's demand for sport programs.

2.2 The Income Side

The income of advertising channels correlates with the rating figures and is generally tied to the total number of viewers. Some channels, however, tie the income to specific targeted groups, for example, the number of viewers between 20 and 45 years old. Thus, information about the factors, which influence the rating figures, is crucial to construct a menu of programs, which maximizes the profit. Furthermore, it is important to bear in mind that the income corresponds with the rating figures during the commercials, not during the programs themselves. Thus, it is also necessary to estimate the correlations between the rating figures during the core program and the commercials.

Maximizing the profit also involves other challenges, such as deciding the duration of the programs. In cases of live sport programs, for example soccer matches, the optimal duration of the program does not necessarily coincide with the length of the match. One alternative is to start the program well ahead of the match and screen a long prematch section filled with interviews, analyses, and similar content. This section can be interrupted by commercials. Such a policy can be profitable if many of those who watch the match also find the prematch section attractive. If not, then the channel can gain from starting the sport program shortly before the match, and instead broadcast other programs up until the match starts.

2.3 The Correlation in Rating Between the Core Program and the Commercials

Even if there are many viewers who follow the match, one cannot take for granted that all of them automatically follow the pre- (and post-) match section. Profit maximizing channels also need information about the rating figures from the nonsport program(s) it can broadcast before the soccer match and compare this with the popularity of the prematch section. The optimal point of time to switch programs is when the net profit contributions from the broadcasting of alternative programs are equal. Finding this point involves analyses of the income as well as the cost side. This article focuses particularly on the income side, that is, factors that influence the rating figures.

Figure 1 illustrates some of the dilemmas that profit maximizing channels face when constructing the menu of programs. Although the figure only concentrates on the period before the soccer match, the principle also applies to the postmatch period. Let us assume that the first program is a nonsport program whereas the second program is a live soccer match. Furthermore, we assume that it is up to the channel to decide when to broadcast the first program, while it cannot influence the time schedule of the soccer match. On the other hand, it does decide the duration of

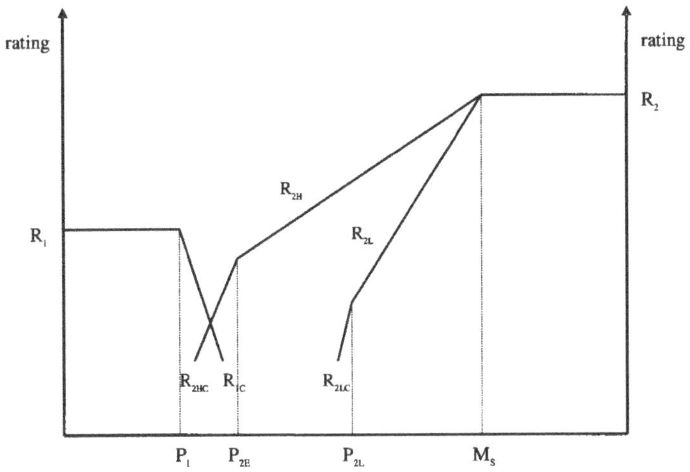

Figure 1. Rating figures during programs and commercials. P = end of Program 1; P = start of prematch program—high rating alternative; P = start of prematch program—low rating alternative; M = start of soccer match; R = rating during Program 1; R = rating during Program 2; R = rating during prematch program—high rating alternative; R = rating during prematch program—low rating alternative; R = rating during commercials before prematch program—high rating alternative; R = rating during commercials before prematch program—low rating alternative.

the soccer match program. As mentioned above, it can start the program well ahead of the match and screen a long prematch section. The alternative is to start the program when the match starts and exclude the prematch section.

R_1 and R_2 represent the rating figures during the two programs. For the sake of simplicity, we assume they are constant over the whole duration of the programs. This is not the case in reality, as the audience during a soccer match can vary considerably. However, the purpose of our analysis is to clarify some principles regarding how to maximize the profit. Having this in mind, such an oversimplification is acceptable. As illustrated, the soccer match is the more popular of the two programs. R_{1C} presents the rating figures during the commercials that are connected to the nonsport program that are screened after this program is finished.

With regard to the prematch section, the figure presents two alternative rating scenarios. The R_{2H} curve refers to a situation in which the prematch program attracts many viewers, while the R_{2L} curve refers to the opposite situation in which the rating figures are significantly lower. The R_{2HC} and R_{2LC} curves present the rating figures during the commercials that are broadcast before the prematch program related to the two alternatives.

Furthermore, we assume that the channel cannot screen commercials during the match. Hence, the entire advertising income will come from the commercials that are screened during the prematch (and postmatch) sections, supplemented by the commercials before the program. The level of this income will be decided by the position of the R_{2HC}—or alternatively the R_{2LC}—curve.

The figure reveals that if the prematch section only attracts moderate interest rating figures, as the R_{2L} curve illustrates, then the channel will gain from starting the soccer program shortly before the match instead of screening a long prematch section. On the other hand, if many viewers find the prematch section attractive, as the R_{2H} curve illustrates, then it will gain from forwarding the nonsport program.

This article concentrates on revealing the position of the curves that are presented in Figure 1. It also investigates to what degree they vary among different categories of matches, keeping in mind that the most popular tournaments are generally sold in bundles. Such bundles can include the entire portfolio of matches or a proportion of them. The terms at which the right fees are agreed on will be influenced by the market conditions. If the competition is tough, the buyers may have to accept bundles that both include less attractive products as well as programs the popularity of which is uncertain.

2.4 Factors That Influence the Ratings

The fact that the popularity of the matches can vary considerably raises some fundamental questions regarding the construction of a menu of programs, which maximizes income. Which categories of matches will draw the largest TV audiences? Is it the matches or is it the teams that will attract most viewers? How loyal are the viewers to their teams? Indeed, a commercial channel can also benefit from not broadcasting each and every match—even if it has acquired the rights. It is possible that the income from other (nonsport) programs may exceed the income from the least popular matches.

It is well-known that some viewers have a preference for specific teams (and competitors/players), such as their own national teams or clubs. Furthermore, some teams have supporters even outside their own geographical region. In soccer, the Brazilian national team is such an example. In basketball, the so-called U.S. dream team, which has participated in the Olympics on some occasions, is another example. Such teams are examples of *profiles*, which is a well-known concept in sport marketing.

In addition there are other variables, which influence the viewers' interest. Some prefer the matches that are of top quality and thus have a preference for the last part of the tournaments where only the best teams remain. So far the research on the demand for team sports has been dominated by analyses of the factors that have influenced the attendances at the arenas, and Dobson and Goddard (2001) and Downward and Dawson (2000) presented an overview of the results from these analyses. Unfortunately, there are few—if any—analyses that have concentrated on analyzing the factors, which influence the numbers of TV viewers.

It is important to bear in mind that TV channels often have limited information about important variables at the point of time they acquire the sports rights. In the case of soccer tournaments, the common procedure has been that the rights are sold a long time before the tournament is played. Attractive products such as UEFA Champions League, the World Cup soccer finals, and the Olympic rights are in general sold several years ahead of the tournaments.

Moreover, some deals run over several years and hence contain many seasons. In such cases even the names of the participating teams or competitors are unknown. This applies to the UEFA Champions League (the source of our empirical data) where the most common duration has been 3 years. Hence, the channels only have limited information at the point of time when they acquire the rights. Moreover, the fact that these tournaments are based partly on the knock-out principle represents an extra risk factor. History contains several examples in which popular teams have been eliminated at an early stage and thus reduced the attractiveness of the tournaments in respective countries.

Consequently it can be more risky to invest in international tournaments than in domestic leagues. Even if the most popular teams may have a bad season and underachieve in the domestic elite league, there is nevertheless no risk that they will be eliminated during the season. Moreover, the channels that acquire the rights to these leagues can give priority to the matches that include the most attractive teams, even if other teams are more successful. Some clubs may find such a policy controversial and in conflict with true sportsmanship. Nevertheless, it gives the channels an opportunity to reduce the risk and hence become less vulnerable. Assuming the channels have risk aversion, such a policy will stimulate them to increase the bids. This will also increase the revenues of the clubs—at least the gross revenues.

3. METHOD

3.1 Data Collection

The analyses presented in this section is based on rating figures from the broadcasting of 116 live soccer matches from UEFA Champions League on Norwegian television in the period from February 2000 to May 2003. All matches were broadcast on TV3, which is a free-to-air channel that is transmitted via cable and satellite. The channel has a penetration of 62% of the Norwegian TV households and is owned by Modern Times Group (MTG). The rating statistics were measured on several occasions during the programs. The majority of the programs started 45 minutes prior to the match, while some started 15 minutes before the match. The programs in the first category contained seven commercial sections, with the first screened just after the start of the program. The following three sections were screened before the start of the first half—with intervals of 8 to 10 minutes. Two sections were screened between the first and second half, while the last one followed immediately after the match. Table 1 provides an overview of the average duration of the various commercial sections and the number of registrations. Some of them started and ended with the screening of sponsor placards. However, this article does not distinguish between the rating figures during these sections and ordinary commercials. The programs that started 15 minutes before the match only had the c4 to c7 variables.

The rating statistics were measured by TNS Gallup Norway in a survey of about 1,000 Norwegian households, equaling 2,400 persons. These households have TV boxes in their homes that are used to register their TV-viewing habits. Each television set is linked to a box that registers whether the TV is on and to which channel it is tuned. All members in the TV household have their own button which registers whether they are watching. The members are recruited from Gallup Norway's annual Consumer and Media survey (http://www.tns-gallup.no/index.asp?tid=

Table 1
Average Duration of Commercial Sections

	C1	C2	C3	C4	C5	C6	C7
Average duration	56	241	229	252	299	301	308
No. of registrations	24	81	86	116	115	115	106

12003&aid=12085&active_color=#319cff&activetitle=&activelabel=Forbruker%20 &%20Media).

3.2 The Profile Dimension

We expected that the viewers had preferences for Rosenborg, the only Norwegian team in the tournament. Thus, their matches will be regarded as one profile category in the following analyses.

Furthermore, Norwegian soccer fans have traditionally been oriented toward English soccer. Over the years, the Norwegian media have paid substantial attention to English teams and players. The Norwegian Broadcasting Corporation (NRK) started to broadcast English elite soccer matches early in the 1970s. Moreover, several English teams have recruited Norwegian players since the early 1990s, which has further stimulated the interest for English soccer. Manchester United and Liverpool have traditionally been the most popular English teams, which is confirmed by the number of their supporters. During the 2000/2001 season, their Norwegian fan clubs had 26,000 and 14,000 members, respectively. This was far above the number of supporters of the other teams: Leeds (3,700), Arsenal (2,300), Tottenham (2,100) and Newcastle (1,000). All the other English clubs had less than 1,000 members (Håndbok, 2001). On the basis of this, we decided to split the matches that included English teams in two categories. So our second category became the matches where either Manchester United or Liverpool were involved, while the third category included the matches, which involved other English teams. The fourth category included the matches which neither included any Norwegian nor any English teams. The distribution of matches was as follows:

- Matches including Rosenborg (22 matches)
- Matches either including Manchester United or Liverpool (44 matches)
- Matches including other English teams (27 matches)
- Matches only including other teams (23 matches)

3.3 The Quality Dimension

UEFA Champions League was launched in 1992 as a reconstruction of the former European Champions cup. The number of teams has gradually been extended—first from 8 to 16 teams, then to 24 teams, and finally to 32 teams. The data refer to a period in which there were 32 teams, and the tournament had three stages. The first stage had eight groups in which 4 teams played a mini-league. Then the 2 best teams in each group advanced to the second stage in which there were again four mini-leagues. From this stage the 2 best teams in each group qualified for the next (and last) stage, the final rounds. Here, the teams first played quarter finals, which were followed by semi finals. The end of the tournament was the final between the 2 best teams. This pattern was changed the season after these data were collected. From the 2003/2004 season, the knock-out procedure will start immediately after the first round.

Due to the elimination process, the teams in the finals should be of highest (average) sporting quality, whereas the teams in the first round should have the lowest (average) quality. Hence, if quality was important to the viewers, one could expect the matches in the final rounds to draw the largest audiences, followed by the matches in the second round, while the first round matches should attract the lowest audiences—all other things being equal.

In this analysis, we tied the quality dimension to the three stages. The basic category was the matches played in the first round, while the matches in the second round and the final rounds were the other two categories. The matches were distributed as follows:

- First round (35 matches)
- Second round (45 matches)
- Final rounds (36 matches—includes quarter finals, semi finals, and finals—in which the winner of the tournament was selected)

3.4 Regression Analysis

A multiple regression model was used to estimate the effects of the independent variables profile and quality on the rating figures in a more precise way than averages figures. We used three profile variables, P_1, P_2, and P_3, respectively, and two dummy variables, Q_1 and Q_2, respectively. As a reference group for the variable profile, we chose Category 4 (teams from nations other than Norway and England). For the variable quality, we used Category 1 (first round) as the reference group. The multiple regression model that relates the rating figures during the matches to profile and quality is as follows:

$$R_M = \beta_0 + \beta_1 P_1 + \beta_2 P_2 + \beta_3 P_3 + \beta_4 Q_1 + \beta_5 Q_2 + \varepsilon,$$

where R_M is the rating during the matches and ε is the error term.

Initially we expected Rosenborg's matches to attract the most viewers that is $\beta_1 > 0$. We also expected that the matches including Manchester United and Liverpool would be more popular than the basic matches, that is $\beta_2 > 0$. The major reason for this is that these two teams have the largest amount of Norwegian supporters.

One could expect that matches including other English teams also should attract larger audiences than non-English teams; due to the attention English soccer traditionally has received in the Norwegian media. On the other hand, the other English teams had considerably fewer supporters than Manchester United and Liverpool. Thus, we had the H_0 hyphotesis that $\beta_3 \neq 0$.

If quality matters, then one could expect the TV audience to increase as the tournament went on. Thus, the finals should have the largest TV ratings, followed by the second round matches, while the first round matches should have the lowest rating figures, in other words that $\beta_5 > \beta_4 > 0$.

Identical regression analyses for each commercial section were also conducted. Here the dependent variable was the rating figures during the commercial section, while the independent variables were identical with the ones presented above.

4. RESULTS

Figure 2 presents the average rating figures during the seven commercial sections and the matches, based on the registrations that were available. As seen, the commercials drew considerably fewer viewers than the matches. In fact, the first section of commercials only represented 18% of the rating figures at the matches. The fourth section, which was screened shortly before the start of the match attracted 37% of the match audience. It was the section that was screened just before the start of the second half that achieved the highest proportion, in fact, with 61%. On average, the rating figures during the commercials achieved 45% of the rating during the matches (see Table 6).

4.1 The Profile Dimension

Table 2 shows the rating figures during the commercials and the matches for the different profiles. We observe that the popularity varied considerably—during the matches as well as during the commercials. Rosenborg's matches attracted the largest audiences, followed by the Manchester United/Liverpool matches. A similar pattern applied to the commercials. However, the data revealed substantial gaps between the two categories. Broadly estimated, the rating figures related to Rosenborg's matches achieved about 100% higher values than the equivalent figures related to the matches of Manchester United and Liverpool. The other English teams, however, attracted fewer viewers than the non-English and non-Norwegian teams. This pattern applied to the matches as well as to the commercials. Indeed the matches only involving non-Norwegian and non-English teams achieved higher figures than English teams other than Manchester United and Liverpool.

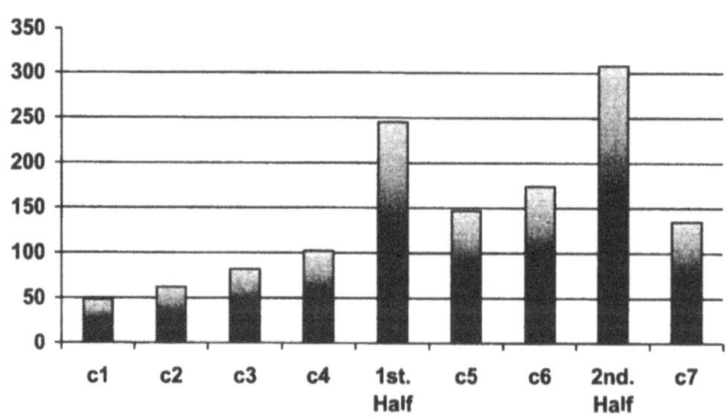

Figure 2. Average rating figures during commercials and matches.

Table 2
Average Rating Figures During Commercials and Matches for Different Profiles

	Commercial No.							
Teams	*1*	*2*	*3*	*4*	*5*	*6*	*7*	*During Match*
Rosenborg	57	98	143	201	275	321	262	469
Manchester United/Liverpool	30	52	73	92	133	158	121	258
Other English teams	48	47	56	64	92	112	91	190
From other nations	58	49	58	68	112	133	98	225
Total	48	62	82	102	147	174	135	276

Note. Figures are given in the thousands.

Table 3
Average Rating Figures During Matches for Different Profiles and Qualities

	Profiles				
Quality	*Rosenborg*	*Manchester United/Liverpool*	*Other English Teams*	*From Other Nations*	*Total*
First round	480	259	177	149	358
Second round	416	234	164	201	222
Final round	—	299	270	235	263
Total	469	258	190	225	276

Note. Figures are given in the thousands.

4.2 The Quality Dimension

The right hand column in Table 3 gives the (paradoxical) impression that the viewers found the matches in the first round to be the most popular ones. The main reason behind this pattern, however, was that all but four of Rosenborg's matches were played in the first round. Over the three seasons, Rosenborg was eliminated in the first round twice, and only once made it to the second round. The columns that do not include Rosenborg's matches reveal a different pattern and indicate that the final round matches were the most popular. However, these figures also reveal that the first matches were more popular than the second round matches—a pattern that will be confirmed by the regression analyses in the following section. This contradicts our initial hypothesis that $\beta_4 > 0$. Exceptions were the matches of non-Norwegian and non-English teams—in which final round matches were the most popular, followed by the second round, while the first round matches were the least popular ones.

4.3 Regression Analyses

To maximize the advertising revenues, it is essential to analyze the variations observed in Figure 2, Table 1, and Table 2. The regression analyses will provide more precise information of how the profile and quality dimensions influence the rating figures both during the matches and the commercials.

The results from the OLS estimation are shown in Table 4. As seen, the R^2 is very high (80%), and so are the F statistics. Hence, the regression model provided a good explanation of the variation of the rating figures. Furthermore, the Durbin Watson observator indicated no significant autocorrelation.

The table confirms that Rosenborg was the most popular team. A typical Rosenborg match attracted 279,000 more viewers than a basic match (not including any Norwegian or English teams). Furthermore, a match that included Manchester United or Liverpool attracted 63,000 more viewers than a basic match. Indeed, the contrasts with the basic matches were significant at one percentage level—both for Rosenborg's matches and for the Manchester United and Liverpool matches. The regression confirmed that the interest in the other English teams was moderate, and that their matches did not attract more viewers than teams from nations other than Norway and England. In summary, the data confirmed that $\beta_1 > 0$ and $\beta_2 > 0$, but not that $\beta_3 > 0$.

The regression also confirmed that the second round matches achieved the lowest rating figures, while the finals were the most popular ones. Hence, the data confirmed that $\beta_5 > 0$, while the hypothesis that $\beta_4 > 0$ was rejected—indeed the regression revealed that it was significantly negative. Although quality seemed to have some influence on the matches' attractiveness, the effect is not as obvious as could have been expected. According to our data, profile seems to be of more importance.

As mentioned above, the channel's revenues will be related to the rating figures during the commercials, not during the matches. Regression analyses for the commercial sections are presented in Table 5. Here the *independent* variables are identical with the ones in the analysis presented in Table 3.

As seen, the commercials related to Rosenborg's matches attracted the highest number of viewers. This particularly applied to the sections that were close to the match and between the first and second half. Regarding Manchester United's and Liverpool's matches, however, only the commercials that were screened close to the match and between the two halves achieved significantly higher rating figures than the basic matches. Although the c1 section achieved lower rating figures, one should be careful in emphasizing this too strongly due to the low number of registrations.

Table 4
OLS-Estimation Results: Match Rating As Dependent Variable

	Coefficients	*SE*	*Standardized Coefficent β*	*t Values*	*VIF*
Constant	194,704	17,784		10.95*	
Rosenborg (β_1)	278,974	19,978	0.97	13.96*	2.57
Manchester United/Liverpool (β_2)	63,062	15,026	0.27	4.20*	2.20
Other English teams (β_3)	1,029	16,516	0.004	0.06	2.04
Second round (β_4)	–26,979	13,543	–0.12	–1.99*	1.80
Finals (β_5)	47,345	15,914	0.19	2.98*	2.27

Note. F test = 85.077**, $R^2 = 0.80$, Durbin Watson = 1.738***.
*$p < .05$, the regression coefficient is significant. **$p < .05$, the model is significant. ***$p < .05$, no significant autocorrelation.

Table 5
OLS Estimation Results: Rating Figures During the Seven Commercial Sections As Dependent Variables

	Commercials						
Teams	*C1*	*C2*	*C3*	*C4*	*C5*	*C6*	*C7*
Constant	46,270	51,743	57,801	83,300	105,146	127,447	89,239
Rosenborg	9,369	44,208	85,201	121,525	173,725	199,333	174,282
t value	0.68	3.53*	6.13*	10.57*	13.11*	13.42*	10.03
Manchester United/Liverpool	–28,516	–3,513	14,728	22,604	34,114	40,774	34,002
t value	–2.5*	–0.34	1.281	2.62*	3.42*	3.65*	2.68
Other English teams	–5,629	9,648	–2,374	–3,910	–3,097	–4	6,615
t value	–0.48	–0.854	–0.19	–0.41	–0.28	0.00	0.47
Second round	7,914	10,496	399	–20,270	23,092	–33,475	–13,686
t value	0.68	1.39	0.05	–2.61*	–2.57*	–3.33*	–1.09
Finals	14,162	–2,451	975	–13,783	17,349	19,487	12,315
t value	1.202	–0.26	0.09	–1.51	1.643	1.65	0.92
R^2 adjusted	0.20	0.38	0.55	0.73	0.78	0.79	0.68

*$p < .05$. (The regression coefficient is significant.)

Furthermore, the analysis partly confirmed that the second round had the lowest audiences. This applied to all commercials just before the start of the match as well as between the first and second half. Regarding the other sections, the rating figures did not differ from the basic matches. It is worth noting that the commercials related to the finals did not attract more viewers than the basic matches, a pattern that contradicted the results in Table 1. Indeed this picture applied to every commercial section. This difference is interesting and confirms that one cannot take for granted that there will be a perfect correlation between the rating figures during the matches and those during the commercials.

Table 6 presents the number of viewers who watched the commercials as a percentage of the rating figures during the matches. As seen, these proportions were higher at Rosenborg's matches, than for the other matches. Hence, not only did Rosenborg's matches attract more viewers, a higher proportion of them also followed the commercials. This pattern particularly applied to the commercials, which were screened shortly before and after the match and between the first and the second half. The differences between Rosenborg's matches and the non-Rosenborg matches regarding commercial numbers four to seven were significant at a one percentage level according to *t* tests. This contradicted the pattern during the first commercial section in which it was Rosenborg's matches that achieved the lowest proportion. However, we chose not to emphasize this difference too much due to the low number of observations.

5. DISCUSSION

This article illustrated that the popularity of soccer matches can vary considerably. As we expected, the TV viewers had preferences for watching their own national

Table 6
Rating Figures During Commercials in Percentages of the Rating Figures During the Matches for Different Profiles

	Commercials						
Teams	*1*	*2*	*3*	*4*	*5*	*6*	*7*
Rosenborg	12	22	31	43	59	68	55
Manchester United/Liverpool	13	19	28	36	51	60	46
Other English teams	24	25	33	35	49	58	49
From other nations	27	21	26	31	50	60	43
Total	18	21	30	37	52	61	48
One way ANOVA *F* statistics	4.7*	1.7	2.0	5.4*	7.0*	5.3*	5.7*

*$p < .05$, at least one significant difference.

team, which was Rosenborg in this case. In addition, many viewers had their own favorites independent of the teams' geographical origin. The fact that Norwegian viewers' had preferences for Manchester United and Liverpool represents such an example. The importance of profiles for people's interest in a sport and/or a competition is a well-known phenomenon in team sport as well as in individual sport. This was illustrated recently by the attention that the sale of David Beckham from Manchester United to Real Madrid achieved in the media. Indeed, this article indicates that the profile of the team influences the rating figures more than the quality of the matches. However, the fact that personalities and profiles receive more attention in the media than content is a well-known phenomenon of the post-modern society, and is definitely not confined to sport (Bauman, 2000).

Although the rating figures indicated that the viewers preferred the finals, other things being equal, they also showed that the matches in the first round were more popular than the matches in the second round. Because 50% of the teams were eliminated from the first to the second round, the second round should be of better quality. Nevertheless, the jump in quality from the first to the second round did not attract more viewers.

The matches in the final rounds were the most popular—other things being equal. However, here it is necessary to bear in mind that the uncertainty of outcome dimension is more important in knock-out procedures than in a mini-league. The consequences of losing (or winning) a match will be more dramatic, which in itself can make the matches more attractive and hence generate higher rating figures. Thus, one should be careful with crediting the higher rating figures in the final matches solely to the quality dimension.

The tendency that profiles are more important than quality can also influence the menu of matches that TV channels offer their viewers. This also applies to domestic leagues. Profit maximizing channels may gain from giving priority to the most popular teams, even in seasons when these are underachieving. Certainly, some teams (and viewers) may find such a policy controversial. Nevertheless, it gives the channel an opportunity to reduce the risk and hence become less vulnerable. Moreover, when the variations in rating figures are as great as in this case, a channel may benefit from not broadcasting all matches. To make the correct decision, it will be necessary to compare the popularity of soccer matches with the popularity of other programs.

As it is the rating figures during the commercials, not during the matches, that decide the income of the channels, the relations between these variables is of vital interest. Although the data revealed some relation, the correlation was not perfect. The commercials that were screened close to the most attractive matches achieved higher rating figures than those for the basic matches, a pattern that also applied to the interval between the first and second half. However, this was not the case for the commercials that were screened a long time prior to the match.

This illustrated that even if a match attracts many viewers, this does not guarantee that the rating figures during the commercials will be as impressive, with reference to the discussion related to Figure 1. It is also worth noting that the commercials related to the final matches did not attract significantly higher rating figures than the basic matches, and that this pattern applied to every section.

When deciding the optimal start (and duration) of the program, the rule of thumb is that the marginal advertising revenues from the alternative programs should be equal. On several occasions, the results indicate that channels might have gained from delaying the start of the programs until shortly before the match—instead of screening a 45-minutes prematch section.

In the cases where the programs started 45 minutes before the match, the rating figures during the immediate commercials achieved only 18% of those viewers who followed the matches. Even the commercials that were screened just before the match did not achieve more than 37% of the rating figures of the matches. This indicates that the optimal strategy is not necessarily to start the program a long time ahead of the match. Here it is interesting to note the difference between TV2's broadcasting of Euro–2000 and TV3's broadcasting of the Champions League. Sixty-four percent of those who watched TV2's Euro-2000 matches also watched the commercials (Solberg, 2004). This was considerably higher than for TV3's UEFA CL matches in which the average was 45%. The main reason for this difference was that the programs from the Champions League matches were considerably longer and included more commercials than the Euro–2000 matches. Although the average Champions League program contained 23 minutes and 30 seconds of commercials, the equivalent figures related to Euro–2000 were 14 minutes and 30 seconds.

As a public service broadcaster, TV2 is obliged to prioritize concerns other than commercial (Solberg & Gratton, 2000). This means that they have to offer a minimum level of specific categories of programs, for example news programs and documentaries. Moreover, they cannot screen as many commercials as the independent advertising channels. However, the data presented in this article indicate that the consequences of these regulations did not damage TV2 severely.

The large variations in rating figures between the different categories of matches illustrate the risk of acquiring the rights to international tournaments. When national teams are eliminated from the tournament, it reduces the rating figures and hence the income in the respective domestic market. On some occasions the national teams do not even qualify. Indeed, this happened the season after these data were collected when no Norwegian teams qualified for the Champions League. It has been calculated that this reduced TV3's advertising revenues by an amount in the region of NOK6 million. Another example is the failure of the Norwegian national team to reach EURO–2004. This can reduce the value of the broadcasting rights to an amount of NOK15 to 20 million, according to Bjørn Gunnar Rosvoll, sales director at TV2, which will broadcast the tournament in Norway (Solberg, 2004).

The examples of unprofitable sports rights deals presented in the introduction also contained several cases of channels that were hit by the lack of success of national teams. This extra risk does not apply to domestic leagues because they do not have any elimination process. Thus, if (too) many channels suffer (too) heavy losses from the broadcasting of international competitions, this can reduce their willingness to bid for such rights. One consequence of this is that commercial channels can prioritize domestic leagues and bid less for international competitions as a way of reducing the risk. Certainly, this will also reduce the fees for the risky products. Another alternative is tying the right fees to the income that the channels receive from broadcasting the programs. For advertising channels, this can be the rating figures, while for pay-TV channels it can be the subscription fees and/or pay-per-view fees. Such clauses will share the risk between the buyers and the sellers. One advantage of this is that it can prevent soccer clubs from being hit by unforeseen financial problems. In some cases the channels are so optimistic regarding the commercial values of the rights that they are unable to pay the price they have agreed. This can cause severe problems if the clubs have entered into contracts with players and other actors believing that income from the TV deals would be stable and high. Indeed, this has been illustrated recently in Germany and England, where many clubs are in deep financial trouble due to the bankruptcies of Kirch Media and ITV–Digital.

On the other hand, such clauses can create a principal–agent problem. The channel (agent) will have more precise information on the variables that decide the fees than the initial owner (principal). This comprises rating figures during the core programs and during commercials, the number of subscribers, pay-per-view fees and also lag effects that occur if viewers on sports programs stay with the channel and continue watching other programs. Such uncertainty can reduce the bids unless a monitoring system is established.

6. CONCLUDING REMARKS

This article analyzed how the profile of the teams and quality of the matches influence the rating figures on the broadcasting of live soccer matches. Of these, the former variable seems to have the strongest influence. Commercial channels will have to balance the incomes from sport programs against the income from alternative programs. The data also revealed that even if soccer matches attract high rating figures, this is no guarantee that the prematch (and postmatch) sections also attract high rating figures. Certainly, there will be some enthusiasts who cannot get enough of soccer and thus enjoy watching the prematch (and after match) sections. Nevertheless, this is not the behavior of the majority of the viewers, according to our data. The median viewer of Champions League matches started to watch when the match was about to start, and thus did not take any interest in the prematch commercials. Hence, the optimal solution for profit maximizing channels may be to start the program shortly before the match.

This article has concentrated particularly on the rating figures during the matches and the commercial sections, which both influence the direct short-term income. We are aware that a TV channel can benefit from emphasizing other aspects than the short-term income. By giving room to a long prematch section, it can build up its reputation as a serious soccer channel. This can make them more attractive among the governing bodies of soccer (and other sellers). Such actors are, in general, interested in maximum publicity for themselves and their sport. Establishing good relationships with them can be regarded as an investment, which can yield long-term payoffs. Next time sports rights are up for sale, the sellers may prefer these channels other things being equal—or alternatively be willing to offer them a discount.

Nevertheless, commercial channels will always have to take into account the short-term financial consequences from the alternative strategies. If there are few viewers who watch the long prematch and postmatch sections, the alternative costs of screening them are high if the channel alternatively could have broadcast other attractive programs.

The results from these analyses should also be relevant to sports other than soccer. There are many international championships and tournaments in which the circumstances are similar to those described in this article. This applies to individual sports as well as team sports. On the other hand, one has to be careful with transferring the conclusions from one sport to another and from one tournament to another. Thus, more research is needed to be able to give recommendations regarding which policy a channel should follow when broadcasting sports other than soccer.

ACKNOWLEDGMENT

We thank Siri Nedrebø at MediaCom for the data on rating figures from the broadcasting of Champions League matches.

REFERENCES

Bauman, Z. (2000). *Savnet fellesskap* [English translation here]. Cappelen Akademisk Forlag.

Desbordes, M. (2003, May). *The relationship between sport and television: The case of the French network TF1 and the World Cup 2002.* Paper presented at the 5th IACE Conference in Neuchatél, Switzerland.

Dobson, S., & Goddard, J. (2001). *The economics of football.* Cambridge, England: Cambridge University Press.

Downward, P., & Dawson, A. (2000). *The economics of professional team sports.* London: Routledge.

Dunnett, P. (1990). *The world television industry—An economic analysis.* London: Routledge.

Fort, R. (2003). *Sports economics.* Upper Saddle River, NJ: Prentice Hall.

Gaustad, T. (2000). The economics of sports programming. *Nordicom Review, 21,* 101–113.

Olson, M. (1971). *The logic of collective action: Public goods and the theory of groups.* Cambridge, MA: Harvard University Press.

Pindyck, R., Rubinfeld, S., & Daniel, L. (2001). *Microeconomics* (5th ed.). Upper Saddle River, NJ: Prentice Hall

Solberg, H. A. (2002). The economics of television sports rights. Europe and the U.S.—A comparative analysis. *Norsk Medietidskrift, 9*(2), 57–80.

Solberg, H. A. (2004). TV-sportsrettigheter—attraktive, men risikable, investeringsobjekter [English translation here]. Retrieved from *http://www.idrottsforum.org/articles/index.html*
Solberg, H. A., & Gratton, C. (2000). The economics of TV-sports rights—The case of European soccer. *European Journal of Sport Management, 7*, 68–98.
Supporterunionen for Brittisk toppfotball [English translation here]. (2001). *Håndbok* 2001.

TRENDS IN COMMUNICATION *12(2)*, 101–116 (2004)

The Fame Game: The Peculiarities of Sports Image Rights in the United Kingdom

Richard Haynes
Department of Film and Media Studies
University of Stirling

A new legal term has entered the lexicon of sport fans. *Image rights*, broadly defined as the commercial appropriation of someone's personality, including indexes of their image, voice, name, and signature, became widely discussed among sports fans after the sale of a leading European soccer player Louis Figo to Real Madrid in 2000. Since then a range of legal, economic, and political arguments have developed in the United Kingdom as to what image rights actually are, their legal efficacy, and their potential impact on developments in the long-standing relationship between sport and the media. In particular, this article focuses on the problematic definition of the term in the United Kingdom context and how it relates to certain economic and commercial transformations in British sport. Using soccer as its primary focus, this article traces the rise of image rights clauses in player contracts. This process is analyzed in the context of rapid and dramatic change in the media coverage of the sport in what is called the new media environment. This article concludes with a range of issues raised by the tightening controls of both proprietary and contractual rights related to sports images and information, and its potential impact on the wider economy and culture of new media sport.

sport, celebrities, image rights, player contracts, soccer, commodification of sport

1. INTRODUCTION

In the Hollywood blockbuster *Jerry McGuire* (1996), the lead character in a fit of moral consciousness decides to write and circulate a new mission statement for his employer that is nonexploitative, ethical, and operates for the "good of the game." McGuire's epiphany rapidly backfires leading to the loss of his job at a major sports marketing firm. Frantically calling his dwindling client list to maintain his presence in the industry he finds himself turning somersaults to secure the contract of a failing running back Rod Tidwell. At the players request McGuire clinches the deal by screaming a mantra down the phone, "Show me the money!" The moment encapsulates one of the broader themes of the film that aims to critique the wheeler dealing of modern sport that is driven by the cult of celebrity and the commercial exploitation of star names. The constant presence of press photographers and broadcast me-

Requests for reprints should be sent to Richard Haynes, Stirling Media Research Institute, Stirling University, Stirling, FK9 4LA, Scotland. E-mail: r.b.haynes@stir.ac.uk

dia that scrutinize the every movement of star athletes combine with the pressures of corporate sponsors and sports organizations eager to turn potentially fleeting success on the field of play into commercial success.

Recent academic research on sports stars (Cashmore, 2002; Whannel, 2002) has also recognized a noticeable shift in the media's obsession with celebrity and the centrality of sport in consumer culture. Such studies point to the symbolic power of sports stars who become bearers of the many aspirations, values, and desires of the fans who follow them. The cultural visibility of sport stars renders them consistently open to media scrutiny and for Whannel, it makes them "points of condensation for a range of discourses" (p. 213) that are constantly reinscribed through the media's coverage of their evolving careers and biographies. Part of this process in the contemporary world of sport is the veneration of stars for their conspicuous display of wealth frequently captured on the celebrity pages of the popular press or showbiz magazines such as *OK*, *Hello*, and *Heat*. Wealth is the adjunct of celebrity and discourses around both structure our understanding of sports stars. Cashmore noted the manner in which "the very existence of celebrities propagandizes the superiority of commodities over things, images over events, and exchange values over worth" (p. 183). Sport, in other words, is transformed from a mere cultural practice involving athletes, players, and teams into a richly mediatized and commodified space for wealth creation. Sport stars are increasingly the cultural leaders of the contemporary urban *nouveau riche* sensationalistically portrayed in ITV's hit British series, *Footballers Wives*.

How the symbolic capital of sports celebrity is transferred in to economic capital becomes interesting in this context. Indeed, as the popularity of programs such as BBC 3's *Celebdaq* and associated celebrity websites that calculate the monetary worth of leading celebrities testify, earning power of major stars is part of the fascination that popular fandom has with them. On any given day in the United Kingdom, sports celebrities may fill as many headlines on the front pages of daily tabloid newspapers as they do on the backpages or sports supplements. Again, as Whannel (2002) asserted, the speed with which information on sports stars circulates in the digital environment drawing on the exchange of stories and images of sports celebrities and media sports events across all media formats comes to dominate the news agenda to the exclusion of all other events. Whannel suggested this creates a "vortex effect" that works to pressurize journalists and broadcasters to concentrate on the celebrity "happening" to the exclusion of all other events. In such an environment images of stars are therefore valuable, both culturally and economically.

For example, in April 2004, Real Madrid and England player David Beckham had his image strewn across the front pages of most U.K. popular newspapers for almost 2 weeks after allegations he had had a string of affairs. The story smashed the public perception of Beckham the devoted husband to celebrity wife Victoria Beckham and doting father to his two children Brooklyn and Romeo. Images of the Beckhams as they attempted to deal with the public relations disaster were at a premium as the paparazzi followed their every move. "Checkbook journalism" also gave handsome rewards to the women who revealed their stories and posed semi-naked to enrapture a voyeuristic British audience. One sobering issue to come out of the affair was the potential damage to Beckham's brand value to the companies he endorsed including Adidas, Pepsi, Marks and Spencer, and Police sunglasses. However, exactly how the market places a value on an individual like Beckham, who earned £53 million in 2003 mainly from commercial endorsements, is difficult to gauge. Creating the legal parameters in which the commercial images from sport can be valued and traded therefore becomes paramount. The claim to image rights is central to this process.

2. MAKING A CLAIM TO IMAGE RIGHTS

Before moving on to discuss some of the key aspects of image rights as they have come to be understood in the British sports industry, it is important to analyze what the key legal and philosophical arguments for a property right in a persons image actually are, and why they have been resisted under U.K. law. *Image rights* usually refer to the commercial appropriation of someone's personality. The term is an imprecise and relatively recent legal nomenclature for indexes of a persons image, voice, name, or signature. Image rights present a new departure in the economics of sport, and the manner in which it is intertwined with the media. In an age in which the cult of celebrity drives much of the media economy, it is not difficult to see why stars and their agents seek to control every aspect of their image. However, as we shall discover, despite the concept of image rights gaining purchase within the sports industry, there is no legislation that recognizes a statuary right of an individual to control the commercial use of their image.

One reason that the English courts have resisted any specific personality rights to protect an individuals characteristics is a general suspicion that property in a person is, in the words of Cornish (1999), "a commodification too far" (p. 31). As we shall see, one consequence of this legal hiatus has been a recourse to a range of intellectual property laws that effectively bolster the ability of leading sport stars to control the use and commercial value of their names and images. This has been deemed highly unsatisfactory for many celebrities eager to prevent others exploiting their fame and notoriety. For example, although some sports stars have successfully applied for trademarks related to their names ("Shearer 9" by Newcastle United striker Alan Shearer) or distinctive images ("Damon Hill eyes" by Formula One driver Damon Hill) trademark law has been reluctant to provide monopoly rights to personalities across all goods and services. There have been well-publicized cases in which third party use of a trademark without permission have not been viewed as an infringement by the courts. For example, in a case involving the legal estate of Elvis Presley the name Elvis was not deemed to be distinctive enough to function as a "badge of origin." Buying a piece of memorabilia with the image of Elvis on it did not instantly suggest to the consumer the source of the product was the Elvis estate (Lane, 2002). Case law had only contemplated damage where there was some common connection in the business activity of the two parties. Quite often monopoly rights in a celebrity image or name cannot be given due to their very popularity in the media and general ubiquitous coverage.

This gap in U.K. legislation contrasts with the United States and other European territories where various personality rights are protected. In the United States the Right to Publicity under the Constitution provides any citizen with the right to control how their identity is used commercially. Crucial in the evolution of the publicity right was the 1953 case involving rival bubble gum manufacturers Haelean Laboratories, Inc. and Topps Chewing Gum, Inc. who were battling to gain the rights to print images of leading major league baseball players on their "trading cards." The Haelean case established that individuals "possessed a property right in their own images" and that "this right could be transferred to a third party" to enforce rights from unauthorized use (Hylton, 2001. p. 274).

The process of commercializing celebrity starts from the belief that one has legitimate proprietary rights in one's own person and any use of ones image, signature, or even style. Debates as to whether sport stars, and any other celebrities in the public eye, should have image rights brings us back to a range of arguments centered in the political philosophy of property. *Intellectual property*, broadly defined, refers to the abstract notion that individuals should have proprietary rights over the creations of their mind. This has been justified in various ways depending on the philosophical point of view. Perhaps the most often cited justification for intellectual property is the Lockean principle of a persons natural right to the fruits of their labor (Drahos, 1996). Applied to the concept of image rights this argument quickly raises some awkward questions. The natural right to own one's image is born of the belief that after years of effort to achieve fame and notoriety individuals who achieve celebrity status have accumulated enough significance for their identity to be recognized as publicly valuable and commercially marketable. Just like a tangible piece of property can increase in value by home improvements, so the reasoning goes, a celebrity's worth as intangible property comes from their efforts to become "somebody" from being a "nobody." The problem in this argument is of course that celebrity, the recognition of an individual in the public domain, is not necessarily the direct result of a person's labor. In the sporting context, labor is expended on the activity of playing sport—of running, jumping, throwing, and kicking—not in turning the success or failure of that activity in to fame and stardom. There are many other factors that create stars, most notably the media and its veneration and celebration of individual performance in sport. As Beverley-Smith (2002) argued, "In most cases, fame—and the opportunities for its exploitation—does not derive directly from the actual process of labouring or from the performance of particular tasks" (p. 295). Indeed, the celebrity images that pervade our contemporary media are part of wider social and cultural processes that represent the focal point of diverse cultural practices and desires. As Coombe (1992) pointed out celebrity names and images

> are not simply marks of identity or simple commodities; they are also cultural texts —floating signifiers that are continually invested with libidinal energies, social longings, and, ... political aspirations. The names and likenesses of the famous are constitutive of our cultural heritage and resonate with meanings that exceed the intentions or the interests of those they identify or resemble. (p. 2)

For sport stars to suggest that their commercial value as a celebrity has been entirely created by the sweat of their brow is both factually misleading and self-deluded hubris. The entourage of agents, publicists, and sports lawyers who act as a vital support mechanism in relationships with advertising agents, sponsors, merchandise licensees, and assorted media outlets is enough to signify that it is not simply sports performance or even hero worship that drives the sport star economy. Rather, it is a whole set of contractual negotiations premised on a perceived value of an individual and their symbolic capital in wider public life.

There are, however, alternative arguments as to why image rights should be protected in law. The utilitarian approach to intellectual property lays emphasis on the ability of proprietary rights to operate as an incentive to innovation and creative production. Under copyright and patent law, authors and inventors are provided an incentive to create, in the knowledge that their works shall be protected in the marketplace to enable some form of remuneration through royalty payments or licences. In respect of image rights, the main problem here is that the lack of a proprietary right in someone's personality, as is the case in the United Kingdom, does not in itself impair the ability of sports stars to exploit their name and image through endorsements or any other commercial activity. Such rights are not needed for sport stars to cash in on their fame.

An alternative argument for image rights under a utilitarian philosophy is the belief that the market for celebrity images and indexes of celebrity rely on restrictive rights to ensure economic efficiency. If a monopoly right is granted to a sport star over the commercial use of her name or image, the argument suggests that the value of access to such a right would maximize its economic worth. In this scenario, an exclusive right to a particular image of a sport star would most likely go to the highest bidder. However, the argument here is also strained in the broader equation of fame and fortune. Maximizing economic value does not always equate with the wider cultural value of celebrity and fame. If images are given away for free or at a marginal cost the coverage of the celebrity is far more likely to be widespread and ubiquitous, therefore more publicly noticeable, conspicuous, and needless to say famous. In some instances winning mass appeal arguably has more potential economic value than exclusively sold proprietary image rights.

The application of mainstream ideas on intellectual property rights to the appropriation of a sport stars image rights are fraught with contradictions. As we shall see through the various case studies below it is not always clear that a proprietary right to an individual's image, name, or likeness is the most effective way of protecting their economic and broader cultural interests. Nevertheless, as Beverley-Smith (2002) suggested:

> At the broadest level, claims for property rights in personality are but one manifestation of the proprietarian creed, which some see as increasingly pervasive in intellectual property law, whereby property rights have a moral priority over other rights and interests, and activities that first give rise to economic value also necessarily create property rights. Property rights trump the wider community interests, and everything is capable of private ownership. (p. 315)

This argument seems particularly apt in the context of sports celebrity and image rights. The increased privatization of star images, allied to a general process of intensified commodification of sport, removes from the debate broader issues of sports culture such as how sport is structured as an aggregated industry and who it is for. In the next section, I address the rise of image rights as a concept in professional sport before moving on to open up broader questions about the cultures of sport and how they are affected by such increased processes of commodification.

In particular, I focus on how this process has been especially pronounced in the football industry. There has been a growing concern of football clubs, players, and their agents with the ownership and commercial exploitation of their media images. I trace the reasons for the increased activity in this area that pulls together a variety of commercial, legal, and political concerns some related to media freedom and others having a deep impact on the relationships between football as cultural industry and its fans. The framing of football fans as consumers runs through the heart of a variety of changes undergone in professional football since the early 1990s (Boyle & Haynes, 2004). The exploitation of player image rights represent a notable staging post in the mediatization and commodification of the sport that is increasingly run (if somewhat blindly) for City investors, sponsors, and advertisers rather than the main consumer base, the fans.

3. PASSING OFF AND SPORT STAR ENDORSEMENT

In the United Kingdom, case law is just beginning to side in favor of sports celebrities and their commercial rights. From a media perspective, the market for celebrity images is certainly a key driver of an increasing will to protect the rights of famous people. Sport stars have increasingly used the media, specifically the popular press with a penchant for celebrity gossip and the dedicated celebrity magazines, to control the representation of their image in an attempt to manage their public perception. Controlled access, such as access to a celebrity wedding, invariably comes at a high cost for the publication concerned and the management of such exclusive access is vehemently policed. When a sports star's image is illegitimately used, various legal remedies have been drawn on to remunerate lack of consent and any resulting injury (financial or to goodwill). The legal representatives of sport stars are turning to innovative interpretations of law to protect celebrity names and likenesses. These include: trademarks as a (a) "badge of origin"; (b) nascent regulations on privacy and Human Rights; and (c) damages brought against goodwill and misrepresentation under the common law tort of "passing off" that have all been used in recent court cases in which well-known personalities have complained about the illegitimate use of their image.

The sports celebrity industry has gained significant encouragement from recent cases involving proprietary rights in sport. First, Formula One driver Eddy Irvine successfully won a court battle against the radio station Talk Radio (now Talksport). Second, the former England cricket captain Ian Botham reached an out-of-court settlement with Diagio owners of the Guinness brand. Third, former British long-distance runner David Bedford threatened to sue the owners of telephone directory service 118 118 after the use of look-a-like runners in the latter's major cult advertising campaign.

3.1 *Eddy Irvine v Talksport*

In March 2002, the High Court decision in the case of *Eddie Irvine v. Talksport* effectively recognized the value of sports image rights and conferred protection on them. Talksport had used a doctored photograph of Irvine in a promotional leaflet that the commercial radio station had circulated to media buyers of potential advertisers. Less than 1,000 leaflets had been distributed but the business affairs director for the Formula One team Jordan, Ian Philips, had noticed the drivers inclusion in the promotion and alerted him to the fact. Talksport had digitally manipulated Irvine's image replacing a photograph of him holding a mobile phone with that of a portable radio carrying the logo of Talk Radio. Irvine's legal representatives argued that the leaflet was suggestive of an endorsement of Talksport by the driver.

The law of "passing off" attempts to protect the goodwill in an unregistered trade mark. The case centered on whether a famous sports celebrity had acquired a valuable reputation and if the goodwill in the stars name or likeness had been misrepresented to the market as being licensed by the individual concerned. Justice Laddie held that the manipulation of the image for commercial gain clearly represented false endorsement. Irvine's value as a brand was viewed as being affected causing damage to his the goodwill, reputation and name. Justice Laddie's ruling confirmed that the name and image of a sports star can be constitutive of a brand, with all the various economic rights associated with that status. More important, it also confirmed that passing off cases involving sport stars no longer rested on the two parties sharing a common field of commercial activity. This was a significant leap in the potential breadth of protection afforded by the courts to celebrity endorsements. Justice Laddie concluded:

> Even without the evidence given at the trial of this action, the court can take judicial notice of the fact that it is common for famous people to exploit their names and images by way of endorsement. They do it not only in their own field of expertise but, depending on the extent of their fame and notoriety, wider a field also...Manufacturers and retailers recognise the realities of the marketplace when they pay for well known personalities to endorse their goods. The law of passing off should do likewise. (as cited in Boyd, 2003, p. 10)

The ruling fell short of providing a complete monopoly right over star images, particularly with reference to character merchandising, but nevertheless the Irvine ruling has encouraged sport stars to believe that the courts will protect their commercial interests when their images and likenesses are illegitimately used.

3.2 *Ian Botham v Guinness*

As English law has made a turn towards recognizing image rights following the Irvine case there have been further instances of potential litigation by sports stars against the misrepresentation of their images and likenesses that were likely to confuse the public in to thinking they had endorsed certain products. The former England cricket captain Ian Botham threatened to sue brewing giant Diageo, the owners of the Guinness brand, when advertising for Guinness used archive images of Botham from the 1981 Ashes series. The Guinness advert juxtaposed photographs from the 1981 Headingley Test match with a tagline quoting another cricket legend Fred Trueman that read, "He couldn't bowl a hoop downhill." The joke being that Botham went on to create cricketing folklore with a brilliant bowling performance.

The advert, run throughout the Summer of 2002 clearly linked Botham's great sporting moment with the Guinness brand. However, Botham's legal representatives claimed that the adverts had used his image without permission. Crucially, Botham argued, his reputation and goodwill were exploited by the campaign and amounted to false endorsement in the eye of the public who might believe the sport star had been paid for the ads. Following *Irvine v Talksport*, Diageo probably recognized the possibility of courts recognizing Botham's claim, and the substantial remuneration he would likely demand for such endorsement, and duly decided to pull the campaign and settle with the former cricketer for an undisclosed sum. The case was widely reported in the national press with the effect that sports image rights gained further recognition and credence in the sports industry and beyond.

3.3 *David Bedford v The Number*

A third, and related case to that of Irvine and Botham, is a claim made by former long-distance runner David Bedford against the telephone directory service, The Number for using a caricature of his persona in their nationwide marketing campaign. The deregulation of telephone directory services in August 2003, has led to a new competitive market for such services previously monopolized by British Telecom. One new entrant, the operator, The Number owned by the US call center company InfoNXX, runs the distinctive 118 118 enquiry service. The service made a major impact on the market in a £70 million cross-media advertising campaign that used a comic theme starring two long-distance runners wearing retro-1970's running vests with two hoops, red socks and pale blue shorts. The characters also had distinctive drooping moustaches and long straggly hair. The ads soon gained a cult following and were critically acclaimed for their humor and innovation. The campaign also appeared to be incredibly successful as the 118 118 service mopped up 50% of the directory service market in the opening months of operation (Oliver, 2003). However, as the advertising campaign gained recognition among audiences and plaudits from the advertising industry Bedford, long retired from athletics and now the race director of the London Marathon, approached lawyers to question whether his image had been misrepresented and illegitimately used for commercial ends.

Bedford made a claim to assert his image rights as the campaign for 118 118 began to gain momentum. In January 2004, Bedford won an important victory when newly formed media regulator Ofcom upheld a ruling that the ad campaign had used a caricature of Bedford that was a breach of Rule 6.5 of the Advertising Standards Code (ITC, 2002) that protects the privacy and exploitation of the individual. The regulator found that the campaign used a "comically, exaggerated representation" of Bedford that was considered against the Code's stipulation that "living people must not be portrayed, caricatured or referred to in advertisements without their permission" (Ofcom, 2004). However, despite recognizing that The Number "neither sought nor obtained David Bedford's permission to be caricatured" the regulator decided that the former runner had not suffered any damages to his reputation or "actual financial harm." Given the amount of money that The Number had spent on the campaign Ofcom ruled that to ban the campaign would be disproportionate to the money already invested by the directory service. The ruling gave a fillip to Bedford's claim that he had not endorsed the campaign but also took any potential sting out of potential civil action regarding loss of commercial revenue from image rights endorsement. Bedford had reportedly estimated damages of £250,000 (Pearce, 2004) based on the ruling in favor of Eddy Irvine and the out of court settlement by Ian Botham.

3.4 Problems With Passing Off Claims

Bedford's case is indicative of the mounting pressure for legislative recognition of image rights and the proactive stance taken by sport stars (even those that have long since faded from the public domain) to control and assert such rights. However, there does appear to be a divergence in the regulations governing the use of sports celebrity images in advertising and endorsements. Under the Advertising Standards Code, Ofcom clearly gave notice that the use of the former athletes likeness should have been sanctioned by Bedford before transmission. However, Ofcom did not have the power to order damages nor did they deem The Number's actions to be punishable by a fine. This decision runs contrary to the judgement in the Irvine case that gave £25,000 in damages after appeal as the amount of money the racing driver would have expected to receive had he endorsed the Talksport campaign. The Irvine case largely rested on the damage to goodwill and reputation of the sport star who was well-known and the public might have expected to endorse such products and services in question. However, as Pearce (2004) pointed out, Bedford was an individual who had fleeting success in the early 1970s and whose reputation as a "celebrity" was virtually insignificant. If anything, the 118 118 campaign had the positive effect of raising awareness of the former athlete putting him back in the public eye after 30 years of obscurity.

An important aspect of passing off cases is the issue of false endorsement. In both the Botham and Bedford cases it is debatable whether the audience would believe that the Guinness and 118 118 ads were commercially connected with either sportsman. In the case of Botham, Guinness would have bought archive material that most people would recognise as being as such. They would not necessarily expect the use of such historical images to be endorsed by the star himself. Unlike the manipulation of the Irvine image by Talksport, Guinness merely used the image of Botham from the 1981 Ashes series as a backdrop to its main message. Similarly with Bedford, the campaign used comic caricatures performed by actors in a parody of a 1970s athlete. Although it was dishonest of The Number to claim that the caricatures were not based on Bedford but rather a "generic seventies athlete," including the U.S. runner Steve Prefontaine, it also stretches the bounds of credibility to suggest that the public would expect Bedford to have implicitly endorsed such a campaign. There was no specific reference that Bedford had endorsed the campaign that would have broken the Advertising Standards Code, and Bedford was not known for endorsing products and would not have the same commercial value as a contemporary sport celebrity.

4. THE RISE OF IMAGE RIGHTS IN BRITISH SOCCER

As case law has moved towards the recognition of image rights certain sport stars have made more proactive arrangements to ensure that permission is always sought for the use of their image. In the context of British soccer, the Europeanization of U.K. player contracts, from the "Bosman ruling" on the freedom of player contracts onwards, has introduced new practices in the protection of star images. The influx of European players in to the English Premiership has educated British born players on the potential windfall gained from "image rights" clauses in their contracts.

As briefly noted above, legislation protecting the publicity rights of sport stars does exist in certain countries of the European Union (EU). In France, legal protection has enabled French sportsmen and women to challenge the unauthorized exploitation of their images and names. French legislation has mirrored the rise of celebrity culture and, under Article 9 of the French Civil Code (http://lexinter.net/ENGLISH/civil_code.htd), is aimed at "protecting the celebrity against the free ride or risks of tarnishment of his of her name" (Boyd, 2003, p. 4). This privacy law has been successfully used to ward off unauthorized use of celebrity images in advertising and other endorsements. For example, former French football star Eric Cantona successfully won a court ruling when his image was illegitimately used on the front of a video. The right under French law is based in the moral right to control one's image and reputation. This cultural difference was emphasized when Cantona entered the English league system and fell in to dispute with his former club Manchester United who had exploited his name to boost sales of replica shirts. Cantona prematurely retired from football to start a new acting career partly in reaction to the over-commercialization of his name and likeness by the Old Trafford club. Subsequent United stars such as Beckham have been made aware of the need to protect themselves from such exploitation and have put image rights clauses in their contracts.

Players in Germany, Spain, and Italy have also turned to the law to guard against unwarranted use of their images and indexes of their persona. For example, German and Bayern Munich goalkeeper Oliver Kahn successfully sued leading computer games manufacturer EA Sports for unlicenced use of his image in the best selling game FIFA 2002. The case was brought despite the fact that EA Sport had signed a rights agreement with FIFPro, the international players union, and the German Bundesliga. However, under German Basic Law (German Bundestag, 2000) individuals are granted commercial rights over their appearance and name. In the German market this law superseded any licence agreement signed by the international players union. An injunction was brought against the sale of the game in German stores that only managed to foster a black market for the game. The case emphasized the zeal with which certain sports stars pursue the control of their image, even where other parts of the industry are collectively licensing goods and services internationally on their behalf. Lack of effective harmonization of image rights legislation across Europe is likely to lead to similar issues being raised as the lucrative market for football related games and related tertiary rights continues to grow.

Despite the growth in such actions, the enforcement of negative rights to protect player images remains relatively rare in European football. Far more apparent is the proactive move by football clubs and governing bodies to reach licensing agreements with players to manage their commercial value and brand. In this respect, U.K. press reports on the importance of image rights in a player's contract first surfaced in July 2000, in the wake of the then record breaking signing of Luis Figo from Spanish rivals Barcelona to Real Madrid for £38 million. A key factor in the deal was a contractual clause over the licensing of image rights, a practice that was repeated by the club in the transfer of Zinadine Zidane, Ronaldo, and subsequently David Beckham. All the deals enabled Real to capitalize on the commercial value of their new acquisitions, owning licence to at least 50% of their image rights. Arguably, the nature of these contracts represents a new phase in the commodification of soccer, with leading European clubs such as Real Madrid at the forefront of this hypercommercialism. As the marketing director of Real Madrid, Jose Angel Sanchez, was quoted: "Image rights will be the fastest-growing revenue line for football clubs in the future" (Boyle & Haynes, 2004, p. 71).

In 2003, aware of the mounting pressure to protect the commercial interests of their members, the Professional Footballers Association (PFA) updated its standard form agreement for player contracts in the Premier League to include a series of agreements that recognizes the promotional role players have under their terms of employment with leading football clubs. Such new arrangements reflect the transformation in the power relations between players and the clubs, and for international representatives, between the players and the Football Association. In 2001, the PFA had reached breaking point regarding the distribution of television money received from the Premier League, Football League, and the Football Association. A small proportion of the rights fees was levied by the union to help fund their benevolence fund, player insurance, and education programs. The dispute centered on the percentage redistributed that historically had been set at 5% of all television revenues received by the governing bodies. However, as the rights deals for 2002 to 2004 were settled, the Premier League offered less than 1% of the £1.1 billion received from BSkyB. The severe reduction in revenue prompted a ballot for industrial action among PFA members. The action would have led to players refusing to play in front of television cameras, which effectively meant all Premier League games would be effected. The PFA received an overwhelming mandate with a 99% from a return of 93% of its members. After a bitter public relations battle between Gordon Taylor, the Chairman of the PFA, and chairmen of the Premier League clubs, a settlement was reached that secured the PFA at least 3% of annual television income, equating to £17.5 million under the current rights agreements. The dispute is interesting in relation to image rights as it brought to the fore issues of performing rights and whether players had any grounds for staking a claim in them. In U.K. copyright law, performing rights are given to musicians and other performing artists but not sportsmen and women. According to Taylor, players have historically foregone their rights in performance "because it is a team game [and, that] they have recognized a particular right and are comfortable not to pursue their individual right with regards to images ... so long as their association has a share" (Boyle & Haynes, 2004, p. 85). It is debatable, however, whether footballers could pursue a right in performance in such a way chiefly because it is a team game and the copyright of a television transmission subsists with the broadcaster and not the player, club, or governing body (Arnold, 2002).

4.1 The Mega-Star Economy

Despite these legal barriers in recognizing proprietary rights in player images the pressure for the insistence on image rights clauses in player contracts has developed from two interrelated processes. The first involves the centrality of stars to the economic success of the English Premier League and the marketing communications practices that have propelled the media coverage of the sport over the past 10 years. From the initial television deal with BSkyB in 1992 originally worth £304 million over 4 years, the latest round of rights has seen their value soar to £1.1 billion over 3 years. This huge influx of cash in to the elite end of the domestic game in England has created a mad scramble for top European, African, and South American talent, to the extent where the majority of players in the Premier League are not from England. The second, and related, process revolves around the ability of Premier League clubs, and those with aspirations to the top flight, to financially resource and sustain the mega-star economy of English football. Clubs have had to find innovative ways of raising finance to attract and pay players often at great financial risk. Image rights clauses have proven a legitimate way of paying a proportion of a players salary under preferential rates of taxation. Players are advised to establish offshore venture capital companies in which image rights payments legitimately earned outside the football domain can be invested beyond the reach of the Inland Revenue. In the late 1990s when the practice first came to light the Revenue investigated the payment of image rights monies to overseas accounts owned by the Arsenal players Dennis Bergkamp and David Platt. The image rights clauses were established to offset the cost of buying the players from Italian Serie A clubs Internationale and Sampdoria, respectively. However, under EU law the transnational flow of money based on nonfootball related commercial income was deemed legitimate.

Some clubs have been driven into a perilous financial state by the hunt for leading talent to bolster their Premier League status and search for a regular place in the UEFA Champions League. In March 2004, Leeds United narrowly escaped administration after accumulating enormous debts of nearly £100 million. Run under a plc, the club had mortgaged future gate receipts through the issue of a security bond raising a £60 million cash injection in to the club. However, the bulk of this money was used on player transfer deals and unsustainable salary levels. When failure to reach the Champions League in 2001/2002 and 2002/2003 led to a significant downturn in television revenues the clubs financial plight spiraled in devastating fashion. Leading players were sold on the cheap to reduce the wage burden and the clubs form slumped as they hit the bottom of the Premier League for the majority of the 2003/2004 season. The financial problems were exacerbated by the nature of some of the player contracts. The club had signed one of its leading strikers Mark Viduka on a hire–purchase agreement with a specialist insurance agency. Viduka also had image rights clauses to bolster his income, tax free, but the insecurity of the plc meant that should the player be sold hardly any of the transfer fee would actually return to the club. At the eleventh-hour a Yorkshire based consortium bought the club for £20 million pounds writing off debt to the major creditors who received 20 pence in the pound. Shareholders in the club, many of them fans who had invested lifetime savings, received nothing as the plc was wound up.

A further indication of the emerging importance of image rights surfaced during the protracted negotiations of David Beckham's contract with Manchester United throughout the 2001/2002 season. The U.K. press followed events with intrigue as Beckham's new found success as captain of England and a rising global icon raised speculation as to his value to the club. Manchester United had embarked on the export of its brand to other parts of the world, most notably southeast Asia and China. The club had also signed a series of lucrative sponsorship deals with leading global brands such as Nike, Vodafone, and Yahoo and Beckham, while a Manchester United player, was central to the promotional activities of the club as it aimed to leverage brand value to attract new communities of fans and generate revenue.

The immense interest shown by the U.K. media throughout Beckham's lengthy negotiations with Manchester United was illustrative of an emergent wider discourse on football finances within the sporting press (Boyle, Dinan, & Morrow, 2002). Image rights also featured strongly in the coverage of contractual talks between Everton and teenage striker Wayne Rooney. Again, the protracted nature of the negotiations stalled not because of the players natural football ability, but more on the value of his image rights. The media's interest in such stories stems from the intersection of a long-standing discourse on players relationships with clubs (will they/won't they sign?), a fascination with fame and wealth, and also a cynicism over the inflated salaries leading players now command. Media coverage of negotiations of this sort frequently chastise the illicit role of players' agents in squeezing more and more money out of the clubs, many of which are increasingly burdened with massive debts (Bower, 2003). However, the key intermediaries in most of these negotiations are specialist sports or media lawyers. Contracts, licensing, and the controlling force of intellectual property are the organizing principles of the football industry. Indeed, the whole exploitation and enforcement of image rights is part of a broader process of the commodification of culture protected under the sign of the law. Increasingly, the sports industries are subject to mounting constraints from the breadth and depth of intellectual property rights and their vehement policing. The regulatory zeal that is sweeping the media and cultural industries in an attempt to shore up any loss of revenue through piracy is also impacting on sport. This enclosure movement has "been broadened to cover more subjects, deepened to cover them for a longer time, widened to cover them in more ways" (Boyle, 2000, p. 2010).

Examples of this process are not hard to find. In May 2003, the Court of Appeal ratified a ruling by the European Court of Justice regarding a trademark dispute between Arsenal FC and the street trader Matthew Reed centered on the sale of "unofficial" merchandise that used the Arsenal crest. Although Arsenal had registered the "Gunners" crest as a trademark in 1989, the traders argument in court was that the crest was in the public domain and acted as a "badge of allegiance" among Arsenal supporters. The High Court initially favored the defendant on the grounds that most people would understand that the traders goods were not official merchandise and did not purport to be so. The ultimate overturning of Justice Laddie's ruling was viewed as a defeat not just of Reed but a whole aspect of football subculture. However, in football's cold commercial age, tradition and sentimentality generally cut no favor.

5. CONCLUSIONS

Despite all this legal maneuvering to protect the commercial worth of players, the term image rights still elicits confusion as to what they are and who is set to benefit.

In an interview with the leading sports lawyer Nic Couchman, he made an important distinction between intellectual property and contractual licences:

> On the one hand there are proprietary rights, and on the other there are contractual rights. A proprietary right is something that basically means I can stop anybody using my property. A contractual right is more about what I am expressly authorizing a third party to do and what I am expressly saying to them what I won't do. (as cited in Boyle & Haynes, 2004, p. 74)

As far as Premier League players are concerned their exploitation of image rights is part of a contractual agreement above and beyond the standard form agreement set out in accordance with the PFA. Despite the fact that image rights are now included in the standard agreement the elite players have far more commercial appeal and as a corollary far more complex contractual negotiations. These individual agreements extend the ability of clubs to exploit a player's image in a group context—such as squad photographs or promotion of club sponsors—to licensing such rights for various commercial purposes exploiting the player's distinctive brand.

From this brief discussion of image rights there are several problematic issues that need to be raised regarding image rights and the new media environment. First, broader issues regarding media coverage of sport and whether information about players, including images and names, is in the public domain are increasingly important. For example, there is often confusion about what can and cannot be placed on the Internet without a licence. It would seem appropriate that where new media content produced by fans is not in direct commercial competition with large media outlets that licences should be waived. At the very least recognition of source material, common within academic writing for instance, would be sufficient acknowledgment of the place of origin of an image or piece of information. One rider on this recommendation is that in the new media environment the distinction between commercial and noncommercial activities begins to collapse. A fan website may carry sponsored links or advertising, or even operate small scale ecommerce activities such as the sale of merchandise, travel clubs, or online auctions for memorabilia. Nevertheless, monopolistic control of images and information would seem to be an overt obstruction on the freedom of expression and innovation the Internet originally brought with it (Lessig, 2001).

The second issue raised is that the globalization of sports commerce has not been matched by appropriate international harmonization of intellectual property law. Despite World Intellectual Property Organisation treaties, established conventions, and European Directives, how image rights are licensed in different territories can cause conflict in the exploitation of sports related rights. The Oliver Khan (Boyle & Haynes, 2004, p. 91) case is just one example, but as the market mobility of leading stars such as Beckham continues apace, problems of illegitimate use of images or more criminal practices such as counterfeiting will evade the global enclosure of intellectual property.

Third, the expansion of certain rights into the digital domain has opened up new economic opportunities for established sports organizations to the detriment of sports media outlets, in particular fan produced media. Since the enactment of the

1998 Rights in Database Act (Data Protection Act, 1998), the football industry, under the auspices of Football Dataco—a joint venture set up by the Premier League and the Football League—has made it an offence to publish any information generated out of the game. For example, all newspapers must pay a license to publish the weekly fixture lists that are deemed a creative database and not a factual list of upcoming events. This has also meant a series of prohibitive costs for fan websites who have traditionally published their teams fixtures as part of the general information fans like to know. It is a classic case of information that was once perceived to be in the public domain now being subject to strict monopoly rights.

Paradoxically, the attempt to commercially control images and other information about soccer is happening at a time of expansion in the communication of all things related to the game. More broadly, legal critics such as Lawrence Lessig (2001) have argued that innovations in new media are potentially being undermined by the enclosure of intellectual property rights. Sport, most notably football and the International Olympic Committee, has been one of the most vociferous industries in protecting domain names and has been involved in a significant number of disputes taken to the World Intellectual Property Organisation and their domain name Arbitration and Mediation Center.

The fourth issue raised by the enclosure of image rights is that it potentially stifles the effective reporting of sport. A case in point was relayed by Niall Sloane, the Head of Football at the BBC in the research conducted for *Football in the New Media Age* (Boyle & Haynes, 2004). He recalled the first time he heard of image rights was when the Brazilian and Middlesborough player Juninho insisted on an image rights fee for a prematch interview with one of his reporters. Sloane telephoned the then Middlesborough manager Brian Robson who authorised the interview and told his player and his agent to stop such nonsense. This suggests that any attempt to monopolize the sporting names and images raises questions of press freedom and potentially threatens freedom of information and expression in the new media age.

Furthermore, new technologies of control are also being introduced to make sure that no unlicenced or illegitimate uses of football's commodity is being literally ripped-off. The implications for a vibrant, creative soccer culture in the digital age can be seen in parallel entertainment industries such as music, in which attempts to ring fence the distribution of digital content have polarized the industry against millions of people downloading MP3 files through peer-to-peer networks.

In the last decade the sports industry has set new legal parameters for its commercial operations. However, although sport can generate huge income in some quarters it is not like any other business. The images and names in sport are not merely economic assets to be exploited, but are embedded more generally in the cultures of sport, from fan driven websites to the posters that adorn the bedrooms of teenage boys and girls up and down the country. Arguably, in the case of British soccer what might be called the cultural contract between football and its fans is changing. Soccer, the oft cited peoples game, is rewriting the bond it has with its fans and this is primarily along commercial lines protected under the sign of law including all the property rights that are generated from the core value of the game, its stars.

Where the early days of sport and the Internet were filled with stories of innovation, creativity, and the multiple voices of fans, today's new media sport is far more concentrated, polarized in favor of established media outlets and driven by global brands. The control of image rights and other forms of intellectual property has much to do with this process that is continuing apace.

REFERENCES

Arnold, R. (2002). Copyright in sporting events and broadcasts or film of sporting events after Norowzian. In E. Barendt et al. (Eds.), *The yearbook of copyright and media law 2002.* Oxford, England: Oxford University Press.

Arsenal Football Club plc v Matthew Reed, EWCA, Civ96 (2003).

Beverley-Smith, H. (2002). *The commercial appropriation of personality.* Cambridge, England: Cambridge University Press.

Boyd, S. (2003). *The legal status of a sportsman's image rights.* Retreived from http://www.selbornechambers.co.uk/site_articles/ Image%20Rights3.doc

Boyle, J. (2000). Analysis, price discrimination and digital intellectual property. *Vanderbilt Law Review, 53,* 6.

Boyle, R., Dinan, W., & Morrow, S. (2002). Doing the business? The newspaper reporting of the business of football. *Journalism, 3,* 149–169.

Boyle, R., & Haynes, R. (2004). *Football in the new media age.* London: Routledge.

Bower, T. (2003). *Broken dreams: Vanity, greed and the souring of British football.* London: Simon and Schuster.

Brooks, J., & Mark, L. (Producers), & Crowe, C. (Director). (1996). *Jerry McGuire* [Motion picture]. Tristar Pictures.

Coombe, R. J. (1992). The celebrity image and cultural identity: Publicity rights and the subaltern politics of gender. *Discourse: Berkley Journal for Theoretical Studies in Media and Culture, 59,* 61.

Cornish, W. (1999). *Intellectual property* (4th ed.). London: Sweet & Maxwell.

Data Protection Act. HMSO. (1998). Available at http://www.hmso.gov.uk/acts1998/19980029.htm

Drahos, P. (1996). *A philosophy of intellectual property.* Aldershot, England: Dartmouth.

Edmund Irvine and Tidswell Ltd v. Talksport Ltd, EWHC 367 (CH) (2002).

German Bundestag. (2000). *Basic law for the Federal Republic of Germany* (Grundgesetz, GG). Retreived from http://www.bundestag.de/htdocs_e/030gg.pdf

Hylton, J. G. (2001). Baseball cards and the birth of the right of publicity: The curious case of Haelan Laboratories v Topps Chewing Gum. *Marquette Sports Law Review, 12,* 261.

ITC. (2002). *The ITC advertising standards code.* Retrived from http://www.ofcom.org.uk/codes_guidelines/broadcasting/tv/advertising_standards

Lane, S. (2002). The problems of personality merchandising in English law—The King, the Princess, and the penguins. In E. Barendt et al. (Eds.), *The yearbook of copyright and media law 2002.* OUP.

Lessig, L. (2001). *The future of ideas: The future of the commons in a connected world.* New York: Random.

McCarthy, J. T., & Anderson, P. M. (2001). Protection of the athletes identity: The right of publicity, endorsements and domain names. *Marquette Sports Law Review, 11,* 195.

Ofcom. (2004). Outcome of appeal by The Number (UK) Ltd regarding complaint by David Bedford. *Advertising Complaints Bulletin.* Retrieved from http://www.ofcom.org.uk/bulletins/adv_comp/content_board/?a=87101

Olliver, M. (2003, October 6). Former athlete to sue over directory ads. *The Guardian.*

Pearce, M. (2004, February 9). An advert or the real thing? *The Media Guardian.*

TRENDS IN COMMUNICATION *12(2)*, 117–130 (2004)

Sport on Commercial Norwegian Radio 1988 to 2003

Peter Dahlén
Ragnhild Thomsen
Department of Information Science and Media Studies
University of Bergen

This article examines trends in sports coverage on commercial Norwegian radio stations. Using historical material, we show how, and why, sport has been given high priority on some of these stations. As the stations commercialize, some reduce the amount of sports coverage, while others increase it. We will explore these somewhat ambiguous trends by focusing on selected stations and by following trends in radio sport to elucidate the general development of commercial radio. We give a short summary of the history of broadcasting in Norway, and make some remarks about radio rights. First, we outline some international research in the area.

broadcasting, deregulation, commercial radio, local radio, formatting, branding, sponsorship, sports media, sports radio

1. INTRODUCTION: RESEARCH ON RADIO SPORT

Although a neglected area of study, some academic studies of radio sport have recently emerged against the background of an ever-developing market. In the United States, where more than 400 million radios receive signals from over 10,000 radio stations sports broadcasting has become a major genre: "Over $2.3 billion is spent on sports radio advertising annually. It is a promotional stable." (Schaaf, 2004, pp. 147–148)

In his study on the history of Notre Dame football broadcasting, *The Fighting Irish on the Air*, Gullifor (2001) presented a key map of the development of radio (and TV) sport in the United States. Gullifor showed how investment increased from the 1980s onwards. In 1982, Notre Dame collected $75,000 in radio revenue, and in 1983 earned its first ever six digit figure from selling radio rights when it secured $550,000 in a 5-year contract with Mutual (Mutual was bought in the mid-1980s and became Westwood One). At the beginning of the 21st century, Mutual/Westwood One pays more than $1 m dollars for the exclusive network radio

Requests for reprints should be sent to Peter Dahlén, Department of Information Science and Media Studies, University of Bergen, Postboks 7800, 5020 Bergen, Norway. E-mail: peter.dahlen@infomedia.uib.no rights to Notre Dame football, more than double what it paid in the mid-1980s (Gullifor, 2001, pp. 68–73, 160).

In an equally groundbreaking and well-documented work, on the play-by-play history of *Sports on New York Radio*, Halberstam (1999) described how sports talk radio was born in New York in 1964 when "the encyclopedic-minded Bill Mazer arrived in New York and introduced the market to its first-ever telephone sports talk show" (pp. 319–320). In the 1970s the play-by-play announcers still enjoyed great prestige in New York, but talk shows were starting to carve a niche. On July 1, 1987, Emmis Communications turned their country formatted WHN into WFAN, the nation's first all-sports station, and, in the words of Halberstam, "the sports radio landscape was forever changed" (pp. 321–322, 378; see Stavitsky, 1995, pp. 88–89). In 1992, WFAN was sold to Infinity Broadcasting, and by 1997, 10 years after it went on air, WFAN was said to reach 1.3 million people a week, with morning host Imus turning over 50% of the audience to the rest of the radio station: "By its 10th anniversary, the all-sports trail blazer was the envy of America's station operators," and since 1987 "more than 50 [radio stations with an all-sports format] have sprouted all over America"—most of them trying to emulate WFAN (Halberstam, 1999, pp. 327–328).

According to Halberstam (1999), in the late 1990s sports talk radio was well established with all-sports radio stations housing different personalities, "some brasher and some softer, some funnier and some more opinionated" (pp. 346–347). An important part of this development occurred when ESPN Radio (run by Disney-owned ABC) was formed in 1992 and broadcast over WFAN on weekends (Halberstam, 1999, pp. 187, 378–379). Catsis (1996) made another major contribution towards mapping the activities and businesses of radio (and TV) sport in the United States. In the United Kingdom, extensive live coverage on football is available both on television and on radio through stations such as BBC Radio 5 Live and BBC Radio Scotland. In addition,

> a substantial quantity of talk-based football content is now available on television and radio. These programs include discussion-based programs like *Any Sporting Questions* on Radio 5 Live, phone-in programs like Radio 5 Live's *606* or Reel Radio's *Reel Football Phone In* as well as comedy-based programs like Radio Scotland's *Off the Ball*. These talk-based programs are also found in many other European countries, most extensively in Italy. (Morrow, 2003, pp. 166–167)

In Scotland, O'Donnell (2003) published a study of another football-based program *Saturday Super Scoreboard*, launched in 1978 on the commercial station Radio Clyde. As one of the most listened-to radio programs in the western part of Scotland, it regularly draws audiences of over 300,000—14% of the population aged four and over in that area. This "reflects an audience pull many television programmes would be envious of." According to O'Donnell, this situation "is by no means restricted to Scotland. So great is radio sports coverage in certain parts of Catalonia, for example, that radio sport there has been termed a 'macrogenre'" (pp. 211–212).

Other studies on radio sport in the United Kingdom include Boyle's (1992) analysis of the link between sport on the radio and broader nation-building processes in Ireland, while Haynes (1999) investigated the origins of the relationship between the Football Association and the BBC in the formative years of 1927 to 1939. A book by the senior football commentator Green (2000), *The Green Line*, on his 25 years in the BBC (on Radio 2 and Radio 5 Live) is also worth mentioning.

In Germany, Gödeke (1976) wrote about the importance of radio sport there during the postwar period. In Sweden, Dahlén (1999) published his dissertation on the Swedish Broadcasting Corporation's radio sports coverage seen in a historical perspective, 1925 to 1995. Dahlén also published on several aspects of early radio sport for youth (2000e), sport in children's educational radio (2001), the popular events around the nationalistic "Vasaloppet" Ski Race in 1952 (2000b), and various other elements of the history of Swedish radio sport (2000a, 2000c, 2000d).

2. NORWEGIAN BROADCASTING—A SHORT HISTORY

The first radio service in Norway started in 1924, and in the following years a number of small, privately owned companies were established in different regions (Syvertsen, 1991, pp. 332–333). In 1933, the public service institution, the Norwegian Broadcasting Corporation (NRK) was established and during the next 60 years, NRK built up a roster of main sport events, some of them inherited from its predecessor the Broadcasting Company (Kringkastingselskapet; cf. Whannel, 1992), and including ski jumping, skiing, and football.

Until television was inaugurated in 1960, NRK had only one radio station. The NRK broadcasting monopoly partly came to an end in 1981 with the deregulation of broadcasting and the advent of local radio, initially on an experimental basis. Advertising was prohibited, by regulation, for all broadcast media in Norway. The local broadcasters therefore had to come up with different solutions to finance thier activity, and bingo, lottery, and "voluntary license fee" were among the sources of income. However it was difficult to draw a clear line between legal and illegal financing, as some types of sponsoring and announcements were allowed despite the advertising ban (Halse & Østby, 2003, p. 197). The development of local radio after the 1981 broadcasting deregulations was rapid, and within a few years numbered 300 stations (Fossum, 1997, pp. 59–60). NRK, moved to address this new competition from local radio and transnational commercial satellite stations. In 1985, NRK opened its second national radio station, P2 (with the original station labeled P1).

The period from 1982 to 1993 was a golden age for sport on NRK Radio, according to former Sports Head Arne Porsum (personal communication, March 19, 2004). With the establishment of a second nationwide station, the sports department got considerable elbowroom to broadcast a greater number of events. The management also understood that sports content was important to attract listeners to other programs in addition to the sports broadcasts themselves.

In 1988, new legislation allowed for permanent local broadcasting and introduced advertising as a source of income. This development was motivated by the emergence of chains of local radio stations, both through illegal trade where stations were bought and sold, usually camouflaged as mergers, and through sophisticated networking arrangements bartering local advertising for national news notwithstanding regulations prescribing local production and ownership.

Through such arrangements the owners of the news source could offer their advertisers de facto near-national coverage. The Legislature was forced to act: "In line with the prevailing liberal ideology, the solution was simply to move the limit and allow a national commercial radio channel." (Fossum, 1997, p. 61) The first private commercial radio channel with national coverage, P4, commenced broadcasting in September 1993. The 10-year license for P4 was allotted to Radio Whole of Norway (Radio Hele Norge, RHN), a company dominated by Swedish-owned Kinnevik.[1] As a political compromise, and in line with the tradition of broadcasting in Norway, public service obligations were imposed. The license terms were vague. The only demands stated explicitly were that P4 had to establish their own news department. However, P4 undertook to build on the fundamentals of public service in the widest sense, emphasizing news, debate, service radio, interaction with the listeners (i.e. phone-ins), "experience radio," and sport (Fossum, 1997, p. 62).

Just 2 weeks after the launch of P4, on October 2, 1993, NRK relaunched its radio services, with three profiled and well-defined stations. The existing channels, the "mainstream" channel (P1), and the "cultural" channel (P2), were supplemented by a third, the "youth" channel Petre (as P3 is pronounced in Norwegian, an acronym for "Program" and "Three"). Petre was launched to recapture the young listeners' segment that had been lost to local radio stations during the 1980s and early 1990s (Skogerbø, 1997, p. 115).

This relaunch was a deliberate response (Fossum, 1997, p. 59). NRK intended to continue being the leading public service broadcaster in Norway. To ensure its hegemony and to meet the mounting challenge from an increasing number of competitors, radical changes were necessary to discard old political regulations and to modernize the institution and its output. The new three-channel division of NRK radio was not without problems, most pressing were the questions of where to place sport and programs in the Samí minority language. A new management relegated sport to the youth channel P3, which only had a coverage area of 90%. Accordingly, in the graduation to three stations and extensive formatting, sport suffered in NRK for some years. In 1996, the sports broadcasts were relocated to P1, the main station, and the conditions improved: "sport had to be on the most popular channel, which is P1" (Fossum, 1997, p. 64). Since then sports coverage has increased.

With the establishment of a nationwide, commercial radio station and a third NRK station in 1993, Norwegian radio stations started formatting their output. Hence, formatting and programming are two central instruments to consider when studying the significance of sport on Norwegian radio stations. Formatting is a phenomenon dating from the 1950s, when the American radio station KOWH–AM found that it could recruit more listeners by restricting the music played to a limited number of popular songs. Since then formatting has been a means to attract a chosen audience segment by use of specialized music profiles. Formatting was an unknown phenomenon in Norway until 1993, since then most radio stations in Norway have been clearly formatted and well-defined.

---

[1]Today Modern Times Group (MTG) dominates (http://www. eierskapstilsynet.no/database/dbprogs/).

3. NATIONWIDE RADIO STATIONS

We now show how, and why, sport has been given high priority on some commercial Norwegian radio stations. The Norwegian radio landscape can be divided into three distinct levels. The highest level is the nationwide stations, and until 2003 this consisted of two: Norsk Rikskringkasting (NRK), which provides several public service stations and is financed by a licence fee, and P4, a commercial, privately owned station. The second level is comprised of chains of local radio stations that together cover large parts of the country.[2] We will compare two key chains, Radio 1 and Jærradiogruppen, and see how these are moving in different directions in terms of sport.[3] At the third level of the Norwegian radio landscape there are around 250 independent, local stations. We will explore a couple of these and show how they use sport in their activities. We stress that the most popular local radio station in Norway, Radio 102 Haugesund, is an independent, commercial station that gives sport high priority.

P4 was established in 1993 as the first nationwide radio station in addition to NRK, as well as the first nationwide, commercial station in Norway. P4 adopted the approved format Adult Contemporary (AC), which has a soft music profile addressed to adults between 25 and 49 years of age, especially women, as its main target group. Radio hosts with easily recognizable personal styles characterize AC-stations, and the audience knows what to expect when tuning in. The program schedule is intended to be inclusive, that is, everybody can listen to it, and the least objectionable music for the target group plays a leading role in nearly all the programs (70% of the broadcast time).

Sport was a priority area in P4's activities from the very beginning, both as part of the editorial content and as a means of branding. From its establishment the station has competed and cooperated with NRK in terms of sport, a dual posture worth noting. P4 and NRK cooperate in securing the rights to broadcast sports events operating the European Broadcasting Union (EBU) policy of open radio access to such events. As a member of EBU, NRK has the rights to major international events, and gives P4 some access. Conversely, P4 owns the radio rights to some important events, such as national and international handball that it shares with NRK. In this respect P4 and NRK are complementary rather than competitor stations. Sports Head Morten Scott Janssen (personal communication, March 23, 2003) for P4 said that his station "always have had a very good dialogue with NRK, and the main reason is that we provide so different products that we don't compete with each other, but we are substitutes. ... We feel that we complement each other."

Thus, P4 and NRK view one another as complementary stations due to different formats and programs. Commercial radio stations such as P4 use formatted program techniques to include and satisfy a wide range of listeners, and the sports broadcasts therefore contain a mix of music, small talk, and short features from the arenas while NRK transmits entire sports events whole. P4 is well defined and clearly formatted. Its rules require that talk is interrupted every 3 minutes by mu-

---

[2]The chains are to some extent profiled as nationwide to advertisers.

[3]*Jærradiogruppen* is the Jær radio group. Jær is short form of Jæren, a region on the West coast of Norway.

sic. This reflects research estimating the audience attention-span at a couple of minutes, and because the station wants to attract all types of listeners particularly women (Morten Janssen, personal communication, March 23, 2003). P4 has worked hard to build credibility regarding the sports coverage. NRK with its long track record was a more trusted sports broadcaster, and the audience distrusted the blended formats. P4 therefore had to change the audiences' listening habits, and persuade them to prefer mixed programs consisting of music and short sports features. Janssen claimed that the audience "has got new habits because they actually don't need to listen to every single kick. They *can* be friends with their wife without missing something. To give them that confidence has been important to us during all these years." Janssen stressed that "our format is holy. We cannot mess with it; it is our most important competitive advantage" (personal communication, March 23, 2003).

Nonetheless P4 and NRK *are* in fact competitors because they both cover most major national and international sports events. Sports Head Grethe Gynild Johnsen (personal communication, February 17, 2004) in NRK commented

> we give one another access to each others' events because we are in principle of the opinion that radio rights should be free for everyone, but we compete in giving these events content .... We have to compete regarding which station fills the broadcast with the best content. Then the listeners have to decide which is the best.

Producing sports broadcasts is expensive; so why does P4 give priority to this kind of content? As a commercial enterprise P4 finds sports coverage profitable. According to Janssen, P4 broadcasts sport because it is a natural theme for a public service broadcaster. In addition, sport is profitable because the audience wants it: "We make radio based on what the listeners really want to hear, and the listeners unquestionably want to listen to sport." Janssen also stressed that "sport is something one can make money on" (personal communication, September 23, 2003). P4 generates some income on advertisements around sports broadcasts, but branding is the primary benefit. In addition to the editorial content, P4 also uses sport through sponsoring Norwegian teams and players. This investment indicates that the station "has sport and uses sport and that you get sport here" (Janssen, personal communication, September 23, 2003). This sponsorship allows P4 to project its brand.

The establishment of P4 in 1993 marginalized the local radio stations, threatening to take both their advertisers and listeners (Halse & Østbye, 2003, p. 245). This has, however, not been the case with sport, and because of relatively different products, P4 has modestly influenced the sports programs in local stations. P4 focuses on nationwide sport, while the local broadcasters offer partisan coverage of local teams. Employees in both P4 and local radio stations tend to regard each other as supplementary, not competitive. Below we show how Norwegian local radio stations have used sport content.

4. CHAINS OF LOCAL RADIO STATIONS

Radio 1 has for many years been the leading chain of local radio stations in Norway. Today it consists of stations in the country's four biggest cities. Radio 1 has a centralized management in Oslo that determines schedules and directions for the entire chain. The stations have, over the years, become clearly formatted and well-defined. They are streamlined and sound more or less the same in Bergen, Oslo, Trondheim, or Stavanger. The radio chain deploys the format Contemporary Hit Radio (CHR) to maximize audience.[4] CHR-stations target the 12 to 34 age group. Content is characterized by a fast-pace, high share of music, small amount of talk, and energetic radio hosts. Contests and branding are important; news has secondary significance. Radio 1 also uses elements from a variant to the AC format, and focuses on women as an important target group.

Sport had been central to the editorial content in at least two of the Radio 1 stations. Coverage has now been minimized with formatting focusing on target groups, marked demands, and profitability. Thus commercialization influenced editorial content. This was particularly evident in Bergen and Trondheim where the local soccer teams in the premier division ("Brann" in Bergen and "Rosenborg" in Trondheim) have a large following. The Radio 1 stations have traditionally covered their away matches; a unique level of coverage. Until 2003, the stations transmitted games in their entirety to a loyal and substantial fan base. A former employee in Radio 1 in Trondheim, Roy Arne Salater (personal communication, January 26, 2004), recalled that almost the entire city listened to Radio 1 coverage. He recalled the priority given to sport: Besides a comprehensive coverage of Rosenborg; the station broadcasted handball, ice hockey, and sports news every morning and evening. It even sent reporters to Brussels to attend the draw of Rosenborg's opponents in the Champions League. The chairman and players of the club were among the audience at home. Advertising for the sports broadcasts sold well, and the soccer reporter achieved celebrity status in the city, a unique achievement in Norwegian radio.

However, the central management decided to cut the sports coverage, reducing the sports broadcasts to a formatted mixture of small talk, music, and short sports features. Some listeners objected to this adjustment; one listener expressed this opinion at Rosenborg's web page:[5]

> Last year Radio 1 gave excellent live-coverage of all of Rosenborg's matches. ... This year however, they have completely changed strategies. ... Now there is a lot of empty small talk and summaries throughout the match interrupted by seemingly endless bits of music.

A year later (2004) a new owner, the international media corporation SBS Broadcasting, removed the remaining sports content. After a gradual reduction, Radio 1 in Bergen is not transmitting sport at all and Radio 1 in Trondheim has terminated its focus on Rosenborg after 13 years of coverage. The competing radio station RadioAdressa has now bought exclusive rights to the club's activities.[6] Interviews with former/present employees of the Radio 1 chain suggest that they do not endorse the management's decision on this.

---

[4]See http://engberg.com/formater.htm.

[5]Rosenborg's Web page is http://www.rbkweb.com/skrivut/1809. The expression quote was posted February 25, 2003.

[6]http://www.propaganda-as.no/php/art.php?id=100785.

In management's view sport is a poor fit in well-defined, formatted stations such as Radio 1. Since 1996, in the formats deployed, sport was a nonpreferred area. Although of interest, it did not command resources within the CHR-format (Pål Thore Krosby, personal communication, January 26, 2004[7]). This format is directed to young, urban, and up-to-date people with considerable purchasing power, and a typical listener is a highly educated woman who enjoys movies, cafés, and shopping. The sports audience did not correspond to this profile principally because they had no loyalty to Radio 1 beyond sport. The removal of sport can therefore be explained as being a fear of losing regular listeners and the recognition that sport does not sufficiently appeal to the target audience. Sportscasting also imports costs of production and rights that greatly exceed those of the single DJ operating from the studio.

Former and current staff in the local Radio 1 stations however disagree with management, stressing the importance of sport in local identification, observing that the chain has dropped one of the more important service, local journalism because of cost. They argue that it does not matter if sport is incompatible with the wishes of the main target group; as long as the sports broadcasts occur outside prime time, that is, evenings and weekends, they will probably attract more listeners than what the radio stations normally could expect at such times. Kim Rud Pettersen (personal communication, January 23, 2004), football reporter at Radio 1 in Trondheim from 1988 to 2001, commented on the decline of sports coverage:

> a majority of the strategies are designed centrally and for the person sitting in Oslo it is hard to see why the people living in Trondheim support Rosenborg .... And they then conceive a collective strategy that is usually made up of music and various humorous features and other content, often at the expense of sporting activities because they envision a nationwide focus, rather than a local focus.

Roy Arne Salater (personal communication, January 26, 2004), a former employee in Radio 1 in Trondheim underlined that:

> it really surprises me that Radio 1 cancelled the sports coverage. Some of the sales staff that worked in Radio 1 were selling advertising like hotcakes to the football matches, so it was very surprising when they suddenly cut that off. ... What Radio 1 has done, is moving away from one of the things I think is the most important of all, that is local journalism. The local broadcasts are a local station's strongest point. I am almost embarrassed on behalf of Radio 1, that has ruled out Rosenborg. They can't even call themselves a local radio station anymore, but perhaps they don't want to do that either. It is easy solutions that can bring in as much money as possible.

Despite dropping sports coverage, Radio 1 still regards sport as valuable in branding terms. As a commercial enterprise Radio 1 postures only towards profit, a policy that is clear in its treatment of sport. Through sponsorship sport remains a means to maintain a local identity. In Bergen, Radio 1 sponsored and supported local indoor soccer tournaments. In return for free advertising, the name Radio 1 is now attached

[7]Pål Thore Krosby was a managing director and editor of Radio 1 from 1996 to 2001.

to the tournament's name, Radio 1 Løkkecup,[8] and its logo is exposed on tournament posters all over Bergen several times a year. Radio 1 in Oslo, is a commercial station that does not transmit sports coverage at all, but nevertheless uses sport in its activities. The station sponsors one of the local premier division soccer teams, and in return gets its logo projected on the players' and coaches' sweaters.

To summarize, Radio 1 is a commercial radio chain that has removed sport as editorial content due to increased formatting and focus on younger target groups. The chain, however, still uses sport as a means of branding. Below we present a commercial chain that moves in the opposite direction in terms of sports broadcasts.

Jærradiogruppen is a chain of 17 local radio stations privately owned by Herman Skretting and A-pressen, a media corporation with business activities within newspapers, printing, TV, and electronic media. The chain emerged in 1997 when Skretting expanded from owning only Jærradioen, a successful local radio station on the West coast of Norway, to become the owner of several stations around the country. In contrast to the Radio 1 chain, the stations belonging to Jærradiogruppen all have a unique identity and can barely be regarded as a unit. This is changing, however. The owners try to streamline the stations and make them sound more similar to compete with P4 for advertisers at the national level.

The stations have already adopted the same target group (13 to 40 year olds), music profile, and style for the radio hosts. Furthermore, the national strategy implies transferring successful programming formats, including sports formats, from one station to the others. A model station in this respect is Jærradioen, the key station in the group, which boasts a very popular sports program concept. Now, we will examine how the station transmits sport and emphasize how it differs from the methods applied by the Radio 1 chain.

Jærradioen is faithful to its sports traditions despite formatting and commercialization. The station still transmits a significant number of sports events, and it uses the format of sports broadcasts that it did in the 1980s. While Radio 1 replaced entire transmissions of events with a mixture of music, small talk, and short features from the arena, Jærradioen still broadcasts entire matches (except for a 2 minute commercial break every quarter of an hour). Sports editor Åge Fjermestad (personal communication, January 23, 2004) in Jærradioen said that "We have held on to a concept since we started in 1987, we haven't changed a thing in the concept. It functions very well. It is dangerous to keep thinking that one should think new. We have a concept that is so secure that we wouldn't change it". Fjermestad emphasized that Jærradioen tries to provide good entertainment by being positive, representing local interest, and it does not have to be "sour, critical and negative like a journalistic angle would be."

Jærradioen values the large audience for sports broadcasts that does not necessarily fit into the target group or return to the frequency the next day. Still they are valued at Jærradioen because they are loyal to the station's sports broadcasts and attract advertisers. Some of the sports listeners also stay tuned to Jærradioen out of loyalty. The sports broadcasts are not profitable in themselves, but enhance the station's profile.

These two chains illustrate how local conditions can determine the emphasis on sports coverage for local radio stations. In the absence of a high profile local team

---

[8]*Løkkecup* is a cup at a local playing field.

sports coverage has very limited appeal for radio stations. Conversely where there are two or more competing high profile teams overt support for one over the other risks alienating listeners. Where there is a single dominant team, with widespread local support, sports coverage provides strong local identification.

We have shown how two commercial radio chains regard sport differently. Now, we focus on independent local radio stations and see how sport can be worthwhile to them in different ways.

5. INDEPENDENT LOCAL RADIO STATIONS

The most popular local radio station in Norway (that with highest number of listeners in its own area) is an independent, commercial station that gives sport high priority. Each news bulletin contains sport, and the station transmits soccer or handball matches every weekend. It might be expected that a station as popular as *Radio 102* Haugesund[9] would adopt the same policy as the Radio 1 chain and offer sports programs as a mix of music, sports features, and small talk to satisfy a wide range of listeners. This is not the case. Radio 102 transmits entire games as Jærradioen does. The station tries to keep a share of 50% music and 50% sport in sports broadcasts, and never interrupts a game with music. Manager and editor Per Helgesen (personal communication, February 12, 2004) said that Radio 102 stakes a lot of money and resources on sport to keep its position in the local media market. The station competes with another radio station as well as with television and newspapers, and regards sport as a competitive advantage aiming to provide better coverage than its competitors.

Their reason for giving priority to sport is that it engages people. Still, Radio 102 is conscious that sport has a limited place in the schedule if it is to avoid alienating listeners who do not take an interest in this type of content. Another reason for broadcasting sports events is that they actually recruit more listeners than what the station could otherwise expect at certain times. Most games and events take place during the weekends, which generally are a poor time for radio. Helgesen (personal communication, February 12, 2004) pointed out that "Radio is a strong medium Monday to Friday, and there are some [listeners] on Saturdays but very few on Sundays. So it [sports broadcasting] is actually about recruiting more listeners than we normally would have on Sundays." It is interesting to note that although Radio 102 evidently is a formatted station, with a music share of 60%, competitions and other types of interactivity with the listeners as well as a high degree of awareness of what the audience wants, the sports broadcasts are not consistent with the rest of the program offering. As already mentioned, this is due to the perception that sport increases the ratings at certain times.

Sports clubs have been among the owners of Radio 102 since its establishment in 1985. A characteristic of Norwegian radio stations is that the local stations were originally established by interest groups spreading ideological and religious information. Due to the advertising ban the owners originally had no commercial motives. The sports clubs and Christian organizations that owned Radio 102 used the radio to inform people about their activities. Gradually, however, Radio 102 be-

---

[9]Haugesund is located on the West coast of Norway.

came a joint-stock company and the founders sold their shares, chiefly to the local newspaper *Haugesunds Avis* that today has a share of 76%. There are still some ideological organizations among the owners, but Radio 102 today regards itself as a commercial enterprise. The goal is to generate income, and their activities started yielding good profits in recent years. To achieve profitability to shareholders, messages were limited and relegated to nonpeak time (Per Helgesen, personal communication, February 12, 2004). As radio has become increasingly professional and commercial, the sports clubs have withdrawn from ownership and from providing editorial content.

Some of the local radio stations continue to be run by religious or ideological organizations that regard broadcasting as a means of spreading a particular message. To recruit listeners, some of them offer a general schedule where sport is among the program genres employed to attract listeners to inform them about a certain viewpoint. *Radio Øst* (Radio East) is one of the stations that regard sport as a way of getting in touch with a wide range of people. Manager Nils Gunnar Haraldseid (personal communication, February 4, 2004) explained that "We were not interested in making a radio just for the Christians. We wanted to make a radio for the public. And of course make them become interested in what Christians stand for."

Radio Øst considers itself as a sort of public service radio, although according to the licensing authorities it is a commercial radio station because it provides paid employment. For this reason we have included ideological stations in an article about trends in sports coverage on Norwegian commercial radio. Profit is not a goal for Radio Øst, and none of its shareholders make money from it. Although other commercial stations cover sport due to their financial motives, Radio Øst has, therefore, purely ideological reasons for the use of sport. The sports coverage is limited to updates of results and achievements, and the extent of coverage is comparable to other current affairs topics (Nils Gunnar Haraldseid, personal communication, February 4, 2004). Up until 1992, however, the station did have a more ambitious sports policy. This included transmissions of entire ice hockey matches with two full-time employees merely covering sport. This commitment was reduced because many of the listeners found the sports coverage to be too extensive.

6. RADIO SPORT BROADCASTING RIGHTS

In the time of broadcasting monopoly (pre-1981) NRK would negotiate with the various sports associations to transmit their activities. In the absence of competition the rates were relatively modest, and because the radio and television negotiated jointly, radio rights were cheap. The question of radio rights received new significance when the Norwegian Parliament inaugurated local radio stations in 1981 ending the broadcasting monopoly. Some of the new stations sought to offer sports coverage but met resistance. For example the Norwegian Football Association (NFF), which controlled rights to the most popular events, feared losing attendance at the arenas if games were transmitted on the radio.

Gradually, the NFF softened its attitude, allowing local radio stations to broadcast sport by granting the individual soccer clubs power to enter into agreements with their local radio stations. NFF only requested to be informed about agreements to prevent local radio stations reselling their rights to other stations. The clubs themselves decided whether the radio stations would pay for the rights in cash or by advertising barter. Some clubs have charged considerable sums, and Jærradioen has since 1990 paid in total more than €118,700 to Viking and Bryne, two top division soccer teams, while Radio 1 in Trondheim paid approximately €11,900 a year to Rosenborg in addition to buying exclusive seats at the arena and promoting the team on the air.

The establishment of P4 in 1993 raised further questions regarding radio rights. NRK held the rights to transmit most major sports events, and P4 therefore had to negotiate with its competitor to get access to these events. As a friendly gesture, NRK allowed the newcomer to participate in their rights. During the 1994 Winter Olympics at Lillehammer, NRK also gave assistance with administrative challenges, studio facilities, and by permitting P4 to transmit parts of NRK's broadcast on its own frequency (Arne Porsum, personal communication, March 19, 2004; Morten Scott Janssen, personal communication, September 23, 2003).

The balance between the nationwide radio stations changed, however, in 1995. As a member of the European Broadcasting Union (EBU), NRK owns the rights to most major international sports events. P4 is not an EBU member, but it gets access to these events through agreements with NRK. This type of resale is a practical arrangement that can be found in many countries, and is approved by the EBU (Grethe Gynild Johnsen, personal communication, February 17, 2004). In 1995, P4 secured some rights, securing exclusive rights to the handball World Championship for women. NRK obtained P4's cooperation to transmit those matches. Women's handball is extremely popular in Norway due to the national team's international success. From this year on, NRK and P4 have alternately exchanged radio rights.

NRK and P4 cooperate first of all because both institutions support EBU's position that radio rights should be available for everyone. This view has an ideological basis of free news communication as well as a practical recognition that technological equipment renders it possible to broadcast secretly (Grethe Gynild Johnsen, personal communication, February 17, 2004). Therefore, the stations do not compete on events, but on offering the best content from the activities. The second reason why NRK and P4 work together is that they do not particularly regard each other as competitors due to the difference in products. Despite the negotiations about access and conditions, P4 and NRK have a stable relationship regarding radio rights. The present agreement lasts until some time in the year 2005, but the stable situation will most likely continue beyond that date.

7. RECENT DEVELOPMENTS

By the beginning of 2004, a second nationwide commercial radio station, Kanal 24,[10] was established and the expansion of the radio field once again raised questions concerning radio sports rights. The issue was whether the newcomer would have access to Norwegian soccer. Some years ago, P4 secured exclusive rights to the com-

[10]*Kanal* 24 means channel 24.

mercial broadcast of soccer, intending to keep future competitors at a distance. Kanal 24 is, however, partly owned by TV2, a privately owned commercial TV station established in 1992 and the television equivalent of P4. This fact was to challenge P4's contract on soccer rights. The media agreement between TV2 and the Norwegian Football Association stipulated that TV2 would also own radio rights if it obtained control of a commercial radio station. The NFF thus had to reconsider the exclusivity given to P4. These circumstances resulted in an ongoing dispute between P4 and NFF that has not yet been resolved.

This dispute about soccer rights illustrates how radio has become an integrated part of television agreements. This threatens P4 because both Kanal 24 and NRK Radio have a close relationship to significant television players, TV2 and NRK TV respectively. P4's future position in one sport is now unclear as TV2 has secured the rights to broadcast handball next year (Morten Scott Janssen, personal communication, September 23, 2003). However even P4 benefits from the rise of all-inclusive media agreements. The station is partly owned by the Swedish media corporation Modern Times Group (MTG) that also owns the television station TV3, distributed via cable and satellite and received by around 65% of the Norwegian population.[11] The TV station owns the broadcasting rights to the Champions League, and as an associate station P4 is able to share in these rights in exchange for advertisements for TV3 (Morten Scott Janssen, personal communication, March 23, 2003).

Because of its short time in operation, the significance of sport on Kanal 24 remains to be seen. Today, the station offers a 4-hr weekly sports magazine every Sunday afternoon; it invests in children's sport by awarding scholarships and it distributes aid to individuals and associations in need of new equipment.

7. CONCLUSIONS

In this article, we have discussed the outcome of the deregulation of Norwegian radio during the 1980s and 1990s, by focusing on sports matters. Alongside the NRK, a public service institution, we have located players at three main levels. For P4, the first private commercial radio station with national coverage, sport is a priority area, while the leading chains of local radio stations at the second level, Radio 1 and Jærradiogruppen, have moved in different directions. Due to formatting policies, Radio 1 has gradually reduced their sports coverage, while Jærradiogruppen continues to give priority to sport, as do, for different reasons, the third level in our study, the independent local radio stations Radio 102, Haugesund, and Radio Øst.

ACKNOWLEDGMENT

We thank Hilde Arntsen for English language editing.

[11]http://www.tv3.no/index.phtml?pagetype=document&document=178&nav=171

REFERENCES

Boyle, R. (1992). From our Gaelic fields: Radio, sport and nation in post-partition Ireland. *Media, Culture and Society, 14,* 623–636.

Catsis, J. R. (1996). *Sports broadcasting.* Chicago: Nelson-Hall.

Dahlén, P. (1999). *Från Vasaloppet till Sportextra. Radiosportens etablering och förgrening 1925-1995.* [From the "Vasaloppet" Ski Race to *Sportextra.* The Establishment and Expansion of Radio Sports *1925–1995*]. Stockholm, Sweden: Stiftelsen Etermedierna i Sverige.

Dahlén, P. (2000a). Att skriva mediehistoria. Den svenska radiosporten ur ett periodiseringsperspektiv. [Writing media history. The Swedish radio sports in a period perspective]. *Nordicom Information* nr 1.

Dahlén, P. (2000b). Karusellmannens underbara öden och äventyr. Radion och Vasaloppet 1952. [The wonderful adventures of the carousel man. Radio and the "Vasaloppet" Ski Race in 1952]. In Bo G. Nilsson (Ed.): *Idrottens själ.* Fataburen 2000, Stockholm, Sweden: Nordiska museet.

Dahlén, P. (2000c). Radiosporten laddar om. [Changing gears at the radio sports desk]. *JMG Granskaren* nr 1, Göteborgs universitet.

Dahlén, P. (2000d). Ska medierna bevaka lokal eller nationell sport? [Should the media cover local or national sports?]. *JMG Granskaren* nr 1. Göteborgs universitet.

Dahlén, P. (2000e). Ungdom, sport och medier. Om populär idrottspropaganda. [Youth, sport and media. On popular sports propaganda]. *Aktuell beteendevetenskaplig idrottsforskning. SVEBIs årsbok 2000,* Lund.

Dahlén, P. (2001). Hälsolära, gymnastik och idrott. Kroppskulturen i skolradion 1930–1939. [Health education, gymnastics and sports. The body culture in educational radio 1930–1939]. *Aktuell beteendevetenskaplig idrottsforskning. SVEBIs årsbok 2001,* Lund.

Fossum, H. (1997). The Norwegian radio reform. In U. Carlsson (Ed.): *Radio research in Denmark, Finland, Norway and Sweden* (vol. 1, pp. 59–69). Gothenburg, Sweden: Nordicom Review.

Green, A. (2000). *The green line. Views from sport's most outspoken commentator.* London: Headline.

Gullifor, P. F. (2001). *The fighting Irish on the air. The history of Notre Dame football broadcasting.* South Bend, IN: Diamond Communications.

Gödeke, P. (1976). *Der Stellenwert des Sports im Hörfunkprogramm: dargestallt an der Entwicklung des Sportfunks bei drei Rundfunkanstalten seit 1945 unter besonderer Berücksichtigung des NWDR (NDR/WDR)* [The validation of sports in radio programming: Exemplified presentation of the development of radio sports on three channels since 1945, with special reference to NWDR]. Münster, Germany: Vlg Regensberg.

Halberstam, D. J. (1999). *Sports on New York radio. A play-by-play history.* Lincolnwood, IL: Masters.

Halse, K. J., & Østbye, H. (2003). *Norsk Kringkastingshistorie* [The history of broadcasting in Norway]. Oslo, Norway: Samlaget.

Haynes, R. (1999). There's many a slip twixt the eye and the lip: An exploratory history of football broadcasts and running commentaries 1927–1939. *International Review of the Sociology of Sports, 32,* 143–156.

Morrow, S. (2003). *The people's game? Football, finance and society.* Hampshire, England: Palgrave Macmillan.

O'Donnell, H. (2002). Fitba crazy? *Saturday super scoreboard* and the dialectics of political debate. In A. Bernstein & N. Blain (Eds.), *Sport, media, culture* (pp. 211–226). London: Cass.

Schaaf, P. (2004). *Sports, Inc. 100 years of sports business.* Amherst, NY: Prometheus.

Skogerbø, E. (1997). Radio for the few or for the many? Experiences from the political history of local radio in Norway. In U. Carlsson (Ed.), *Radio research in Denmark, Finland, Norway and Sweden* (pp. 107–121). Gothenburg, Sweden: Nordicom Review.

Stavitsky, A. (1995). Ear on America. In E. C. Pease & E. E. Dennis (Eds.), *Radio: The forgotten medium.* New Brunswick, NJ: Transaction.

Syvertsen, T. (1991). "Culture" vs. "business:" Structural changes in Norwegian broadcasting. In H. Rønning & K. Lundby (Eds.), *Media and communication* (pp. 331–344). Oslo: Norwegian University Press.

Whannel, G. (1992). *Fields in vision. Television sport and cultural transformation.* London: Routledge.

TRENDS IN COMMUNICATION *12(2)*, 131–152 (2004)

Tackling the Directive: Television Without Frontiers, Transnational Broadcasting, and Irish Soccer

Roderick Flynn
School of Communications
Dublin City University

With the liberalization of European Broadcasting from the 1980s and the early identification by new commercial channels of sporting events as key content to encourage viewing/subscription, the relationship between sports and media organizations has inevitably become closer and more lucrative. However with the amendment of the European Union's (EU) Television Without Frontiers Directive in 1997 a new element—the state—entered the equation, rendering this relationship all the more complex. This article demonstrates the extent to which, in the modern era, sports journalism can only offer a truly comprehensive account of events by adopting approaches more commonly associated with business and political journalism. It does so by examining the biggest Irish "sports" story of 2002: not the Irish teams participation in the World Cup but the subsequent sale of broadcast rights for that team's home games to BSkyB. The narrative throws light on the increasingly complex relationship between sport, commercial, and public service media, and the state but also details an extremely novel application of the Television Without Frontiers Directive by the Irish state.

broadcasting regulation, Television Without Frontiers Directive, European Union, special events, Republic of Ireland, football, BSkyB, journalism

1. INTRODUCTION

On July 5, 2002, 3 weeks after the Republic of Ireland football team returned from the World Cup in Asia, the Football Association of Ireland (FAI) sold to BSkyB the exclusive live television rights to Republic of Ireland home internationals for the next 4 years. The deferred coverage rights had additionally been sold to TV3, Ireland's only privately owned[1] free-to-air commercial broadcaster. The deal would immediately affect Ireland's upcoming round of European Championship qualifying matches.

Requests for reprints should be sent to Roderick Flynn, School of Communications, Dublin City University, Dublin 9, Ireland. E-mail: roderick.flynn@dcu.ie

[1]TV3 is jointly owned by Can West Global (45%) and Granada–Carlton (45%).

The Irish public response was outrage: hitherto such matches had been available on a live free-to-air basis via RTÉ, the Irish public service broadcaster. Now fewer than one in six homes—those with access to BSkyB's subscription services—would be able to see such matches. Parents without Sky subscriptions complained that their children would have to go to pubs to watch the games (as many Irish public houses offer access to Sky's Sports Channels). Depending on your perspective, access to a key element of national culture had been sold for a "mess of potage" by the FAI, or stolen by BSkyB. Public opprobrium was poured on both organizations, although the FAI was particularly singled out for the underhand way in which it was perceived to have negotiated the deal. On the defensive, the FAI in turn blamed RTÉ, noting that at a time when the organization was seeking to develop the game in Ireland, the €7.5m offered by Sky far exceeded the €1.6m offered by the state broadcaster (O'Sullivan, 2002). Indeed the FAI pointed out that RTÉ's offer was lower than the amount the public service broadcaster had paid for the previous 4 years' matches. RTÉ in turn defended their position by arguing that the value of sports rights was falling internationally, with the then recent collapse of the OnDigital deal with the Nationwide League in the United Kingdom, originally worth £315m replaced with an offer from Sky of £90m. Meanwhile several government ministers including the Minster for Communications, the Minister for Sport, and ultimately the Irish Prime Minister, all expressed their disappointment at the deal but glumly averred that the state had no means of intervening in the situation.

In passing, each of the politicians acknowledged that the 1997 amendments to the European Union's "Television Without Frontiers" directive (European Council, 1997) empowered EU member states to draw up a list of events, which had to be made available on a free-to-air basis. Furthermore they acknowledged that these powers—granted under Section 3a of the directive—had been transposed into Irish law 3 years earlier via the Broadcasting (Major Events Television Coverage) Act 1999. Nonetheless they stressed that despite strenuous negotiations with a range of sporting bodies representing soccer, rugby, and Gaelic games over the previous 3 years they had been unable to come up with a list satisfactory to all the parties concerned.

And that apparently was that—the Football Association of Ireland clearly anticipated that having announced the deal late on a Friday afternoon that it would have blown over by after the weekend. The story did not disappear, reflecting two things: first that games involving the national team were widely regarded as events binding the country together. As one commentator put it:

> The Irish home internationals, while not as popular as World Cup games, are still the kind of events that draw whole families and communities around their television sets. In the atomized media age in which we live, there are not many events like that any longer. (Oliver, 2002, p. 5)

Second, the story lent itself to coverage from a range of perspectives reflecting the different constituencies represented by the various institutions involved. These institutions were the FAI, RTÉ, BSkyB, and the Irish State. To understand why the story "grew legs" but also the complexity involved in properly covering it, we need to briefly outline each institution's role and perspective, and also the role played by the Television Without Frontiers Directive in domestic broadcasting regulation.

2. TELEVISION WITHOUT FRONTIERS

The main intention of the original 1989 version of the TWF Directive was to facilitate the creation of a single European market in audiovisual trade. The preamble to the directive stressed that:

> It is essential for Member States to ensure the prevention of any acts, which may prove detrimental to freedom of movement and trade in television programmes. (European Council, 1989, p. 24)

In the original 1989 version of the directive the third article forms two short paragraphs, which qualify stipulations laid down in the directive's other articles. Clearly reflecting the principle of subsidiarity, it permits individual member states to "lay down more detailed or stricter rules" than those embodied in the directive (European Council, 1989, p. 26).

As amended in 1997, however, Article 3 becomes much more substantial. It also creates an exception to the directive's attempt to do away with barriers to trade in audiovisual products. The amended Article 3a states that:

> Each Member State may take measures in accordance with Community law to ensure that broadcasters under its jurisdiction do not broadcast on an exclusive basis events which are regarded by that Member State as being of major importance for society in such a way as to deprive a substantial proportion of the public in that Member State of the possibility of following such events via live coverage or deferred coverage on free television. (European Council, 1997, p. 66)

In effect the article permits Member States to insist that specific cultural or sporting events be broadcast to their citizens on a free-to-air basis. Furthermore it enjoined Member States to ensure that broadcasters in their jurisdiction did not exercise exclusive rights so as to deprive the public in another Member State of the possibility of watching events, designated by that other Member State to a free-to-air basis.

In passing it is worth noting that the directive thus embodies the European Union's occasionally schizophrenic attitude towards the audiovisual industry (albeit one justified by the dual nature of the industry). Whereas the main intent of the directive is to remove artificial obstacles to audiovisual trade within the EU, Article 3a echoes the line consistently adopted by the EU since the 1993 GATT negotiations: that cultural products cannot simply be treated as commodities.

The remainder of the article outlines the mechanics of how such events could be protected. Member States were to notify the Commission of any list of events being drawn up. In turn the Commission was obliged to seek the opinion of another innovation introduced by the 1997 amendment—a Contact Committee constituted by "competent authorities of the Member States"—as to the appropriateness of the list. Finally, the Commission was given 3 months in which to inform the relevant

Member State as to whether the list was acceptable or not. As of 2002, five countries had submitted lists to the Commission: Germany, Italy, the United Kingdom, Austria, and Denmark and a further three (The Netherlands, Belgium, and France) had notified the Commission of their intention to adopt similar lists (Commission of the European Communities [CEC], 2003).

The size and variety of lists varied: The United Kingdom listed 19 separate events while Germany listed just 5. Virtually all of the lists included the Olympics and soccer's World Cup/European Championships. Indeed the bulk of the events listed were of a sporting nature, although in Italy and Austria several musical events were also included.

Until 2002 the application of Article 3a in individual Member States had been relatively straightforward. Indeed the only substantive block to the Article's operation came at the behest of a Member State rather than the Commission: on January 1, 2002 the new conservative Danish government revoked the list drawn up by their socialist predecessors on the grounds that they "considered that the arrangement in question was not compatible with free competition on the market" (Ministry of Culture–Denmark, 2003). This then is the European legislative context in which the sale of Irish television rights took place. What then was the role of the other key players?

3. THE FOOTBALL ASSOCIATION OF IRELAND

Prior to 1988 the Republic of Ireland had never qualified for a major international soccer championship. However after qualifying for the 1988 European Championships the team went on to qualify for the 1990, 1994, and 2002 World Cups. The net effect of this was that the FAI, an organization run in part by volunteers or part-time staff, was forced to professionalize its own activities, a process not without its growing pains—the FAI was roundly criticized for its failure to capitalize financially from the team's involvement in the 1990 and 1994 World Cups. Nonetheless as the 1990s progressed lessons were learned, the FAI would net in the region of €1m from the Republic of Ireland's involvement in the 2002 World Cup (Dooley, 2002).

This professionalism was also evident in the manner in which the FAI adapted to a changing international market for sports rights. In September 1994, the FAI signed a 4-year deal with RTÉ, which renewed what was by then regarded as the station's customary coverage of the Republic's international home games. The value of the 1994 deal was never published but according to a joint FAI/RTÉ statement the sum involved was "a significant increase on the previous sum, and substantial" (Thornley, 1994, p. 19). This deal was further eclipsed by another agreement in 1996, which at the time was the biggest in the history of the RTÉ Sports Department. Nonetheless the sums involved in any of RTÉ's deals with the FAI would clearly be eclipsed by the BSkyB deal in 2002. From a commercial perspective the FAI had apparently been extremely successful, given that it had secured an offer of €7.5m from BSkyB for its rights in comparison with the €1.2m offered by RTÉ, the use of the latter offer as a means of leveraging €7.5m from BSkyB being quite an achievement.

The scale of the negative public response to the BSkyB deal makes it clear that the professionalism exhibited in negotiations had not yet extended to the FAI's

Public Relations operations. In contrast to the PR savvy demonstrated in the English and Scottish professional leagues (where many of the Republic of Ireland's players ply their trade), the FAI has, at least until recently a history of scoring PR own goals. Although Boyle, Dinan, and Morrow (2002) noted, with regard to the English and Scottish Football leagues that the closed world of football journalism is "perhaps beginning to be transformed as the influence of agents and PR increases" (p. 167), this influence was in its infancy within the FAI at the time of the sale of the rights relating to the Irish team.[2]

In consequence the release of the news about the television deal was not handled well. The timing was appalling: on June 16, 773,000 viewers had watched the team suffer a glorious defeat at the hands of the Spanish team and exit the World Cup in its second round. To welcome the team home a few days later, 100,000 people turned up in Dublin's Phoenix Park . These events were still fresh in the public mind when the FAI announced the BSkyB deal 3 weeks later. Furthermore, the FAI's initial announcement made scant effort to place a positive spin on the fact that 4 years worth of games were about to become unavailable to the majority of the population: rather it adopted a petulant tone, blaming the loss of the games on RTÉ's failure to come up with a realistic offer.

A more positive spin was offered later but only in response to a week of public outrage expressed through the letters pages of the national press and radio phone-in shows. In consequence, the FAI's more nuanced justification for the deal—which focused on the need to invest in schools soccer, upgrading the national coaching system and investing in facilities and infrastructure around the country—largely fell on deaf ears. In effect, in the furore over the BSkyB deal, the fact that the FAI had acted not merely as a professional organization but also as a business, was lost.

Also lost in the furore, however, was the FAI's reading of the state's failure to designate the home internationals as per Article 3a of TWF (or Article 4 of the Irish legislation that transposed the article into domestic law) in the 5 years since the Directive had been amended. Quite reasonably, the FAI took this to mean that they could enter negotiations with organizations like BSkyB without political interference. This was not to be the case.

4. THE POSITION OF THE PUBLIC SERVICE BROADCASTER; RTÉ

The reaction of the Irish public service broadcaster to the BSkyB deal was to seek to portray itself as the victim of bad faith. RTÉ asserted that the FAI had signed the BSkyB deal whilst still nominally negotiating with RTÉ (an assertion refuted by the FAI). However, reading between the lines, it appeared that the broadcaster was also

[2]For example in the 12 months leading up to June 2002, the FAI had been criticized in the press for its involvement in an abortive (but expensive) attempt to establish a new national soccer stadium. Later, when the star player of the Republic of Ireland side was sent home from the Irish team's Saipan World Cup training camp before the team's first match following a highly public dispute with the manager, somehow the FAI again took the brunt of the blame. In the wake of what became known as "the Saipan meltdown," the organization had belatedly employed the services of an independent PR consultant (see Humphries, 2003).

reeling from shock: There was a sense that the station had operated under the assumption that such a loss simply would not happen.

Such confidence is reflected in the specific terms of earlier deals between RTÉ and the FAI. The 1994 RTÉ/FAI deal was due to run until 1998 but only guaranteed the station coverage of one competitive game in either the World Cup or European Championship qualifying series. In practice RTÉ had covered virtually every game in the qualifying rounds of the respective 1994 and 1996 World and European championships and it appears that both RTÉ and the FAI had simply assumed that this would be the case. The casualness of the deal's terms appeared to reflect RTÉ's confidence that its de facto status as a television broadcasting monopoly meant that a deal with more specific terms was unnecessary. In some respects the relationship paralleled that which existed between the English FA and the BBC/ITV duopoly prior to the establishment of BSkyB: when the English League attempted to dictate terms relating to television rights at the beginning of the 1985/86 season, both the BBC and ITV simply ceased broadcasting live matches until the League—suitably chastened—returned to the negotiating table (Baimbridge, Camerson, & Dawson, 1996). Similarly, in 1994, RTÉ was apparently able to dictate terms to the FAI, a fact that must have chafed with the latter organization.

However, if in 1994 RTÉ were apparently immune to the threat to broadcast sports rights posed by corporate interests such as the Kirch group and BskyB, the same was not true 2 years later. Midway through the 1994 to 1998 deal, motivated in part[3] by speculation that BSkyB was considering bidding to secure exclusive broadcasting rights to all Ireland's home games, RTÉ re-entered negotiations with the FAI. The result was a deal—the biggest hitherto negotiated by RTÉ Sport—entitling RTÉ to cover all home games played by the Republic of Ireland, competitive or otherwise, until 2002 (Byrne, 1996).

Between 1996 and 2002 however, RTÉ, like public service broadcasters across Europe, continued to see new commercial competitors emerge. By European standards RTÉ's status as a television monopoly had been unusually long-lived. Although legislation ending the station's state-mandated monopoly in television broadcasting was passed in 1988, it was not until 1998, that the first private television station—TV3—took to the air. Given the clear and present danger posed by TV3 and BSkyB, it seems surprising that RTÉ did not take the threat of competition within the Irish market more seriously when negotiating with the FAI in 2002.

However, the station was financially ill equipped to face such competition. In the 1990s RTÉ not only faced more competition but also competition of a different nature than previously. In theory, RTÉ had operated in a competitive environment since its inception in 1961 as viewers living on the east coast of Ireland already had access to both the BBC and ITV output. As new channels came into being in the United Kingdom in the following decades they have almost automatically become available to the Irish audience. This tacit competition with U.K. channels for audiences did not extend to competition for advertising revenues, however: RTÉ retained a de facto monopoly in the Irish television advertising market until the

---

[3]The renegotiation may also have reflected the first attempt by RTÉ's long-time head of sport Tim O'Connor, to retire. O'Connor did ultimately retire in 1999/2000.

1990s. However, as competition within the U.K. market in particular heated up in the early 1990s—with the arrival of Sky and Channel 5—the established U.K. terrestrial stations began to consider new markets. Hence in 1993, both UTV[4] and Channel 4 began canvassing for advertising from Irish companies (Foley, 1993). As of early 2004, MTV, Sky One, E4, and Nickelodeon were also offering Irish advertisers local opt-outs during ad breaks to allowing them to target advertising to Irish audiences.

The net effect of this on RTÉ has been dramatic. Not only has the station lost market share[5] but it is also losing advertisers. In theory this should not be an overriding concern for RTÉ: As a public service broadcaster the station could expect to be at least partially shielded from the demands of the market by the license fee. However, at the time of the BSkyB deal successive government refusals to countenance a license fee increase meant that the Irish television license was one of the cheapest in Europe: This had forced the station to look to commercial sources for funding. Thus by 2002, 70% of the station's annual income came from commercial sources (Flynn, 2002a, p. 170). Even with this the station was still losing money—in 2001 it recorded a deficit of €70m (Radio Telifís Éireann Authority, 2002, p. 50).

Ironically, precisely because the station was poorly financially equipped to engage in a rights bidding war, its reliance on commercial income meant that the games were important not just to RTÉ's cultural mission as a public service broadcaster but also because of their potential advertising revenues. The average TVR for Ireland's World Cup qualifying games between September 2000 and October 2001 had been a healthy 13.7, despite the fact that several of the games took place during work hours against competition (Cyprus & Andorra) who wouldn't normally attract an audience. Thus the loss of the games to BSkyB was a double blow.

5. SKY'S POSITION

The position of BSkyB in the Irish affair is inevitably a complex one given its place in the intricate cross-media empire that is News International (NI). NI owns 36% of BSkyB. It also owns the *Sun*, the *Sunday Times*, and the *News of the World*, three British newspapers that enjoy substantial sales in the Irish Republic.

BSkyB operates on a number of levels in Irish Television. At one level it is a simply a television channel: 57% of Irish television households have access to Sky One and Sky News, the station's unencrypted channels, although that figure rises to 91% in the Dublin region. Indeed such is their interest in the Irish market that in Spring 2004 the station announced the production of a dedicated Irish news program (Seoige, 2004). Meanwhile the station's three subscription sports channels enjoy household penetration rates of between 15% and 19%.

BSkyB is also an important television service provider in Ireland—as of December 2003, 297,000 Irish homes subscribed to Sky's digital satellite service. This represents a remarkable success story for the company. Having only launched digital

---

[4]Holder of the ITV franchise for Northern Ireland.

[5]In 1997 RTÉ's market share amongst those viewers with access to both U.K. and Irish stations (termed *multichannel viewers* in Ireland) was 45%. By February 2004, this figure had fallen to 39.1% (see A.C. Neilsen, 1997; 2004).

television in Ireland in 1999, Sky's subscriber base now accounts for more than 20% of all Irish homes and 75% of the domestic digital market (the rest being mainly accounted for by NTL's digital cable service; Commission for Communications Regulation, 2003, p. 17). Hence BSkyB is the dominant player in the Irish digital market, especially given the collapse of a government strategy to introduce Digital Terrestrial Television.

The company's digital strategy is reliant on attractive local content. Rupert Murdoch's identification of sport and movies as the battering ram that would gain entry for his channels into people's homes has been repeated to the point of banality but it remains a key element in the strategic plans of his television interests. Thus although one doesn't need digital television to access the Sky Sports Channels, the manner in which Sky structures its various suites of channel offerings use the content of the sports channels to encourage sports channel subscribers into installing Sky digital. The "ordinary" price for a Sky digital installation is €100 but if one opts for Sky's premium package (which includes all three sports channels plus two movie channels) that fee is dropped to an almost negligible €15.

This explains BSkyB's interest in Irish soccer. The channel had in any case already become somewhat unwillingly involved in funding Irish football some 6 years earlier. Under UEFA rules, in agreeing broadcast rights deals, a given country's football association must have the permission of other associations potentially affected by that coverage. Thus when the English FA sold the Premiership screening rights to BSkyB in 1992 in a deal that envisaged screening a hitherto unprecedented[6] 60 live matches each year, they had to discuss the impact with the FAI. Because Premiership matches were widely watched in Ireland, the FAI were entitled to compensation for lost crowds at Irish games. Such payments could not be made in the form of cash, however: thus in the early 1990s BSkyB put £IR1.5m into upgrading facilities at Irish grounds. Furthermore, in 1999, BSkyB had purchased rights to screen the Republic of Ireland's home matches on U.K. television, in an attempt to appeal to the substantial Irish community living there.

As such the €7.5m offered by BSkyB made doubly good business sense: given the cost in 2002 of subscribing to Sky's Sports channels and the fact that the annual cost to BSkyB of the deal was €1.875m, Sky would only need an additional 3,000 to 4,000 subscribers to offset the cost of the deal. At a point when Sky had already secured 200,000 plus Irish subscribers, a further 3,000 to 4,000 seemed an entirely likely prospect.

6. THE POSITION OF THE IRISH GOVERNMENT

Of all the actors in this narrative the rationale for action (or indeed inaction) of the Irish government is most opaque. It is important to stress that the same Fianna Fáil/Progressive Democrat coalition in power at the time of the BSkyB deal had 3 years earlier introduced the Broadcasting (Major Television Events Coverage) Bill, which transposed Article 3a of Television Without Frontiers into Irish law. Intro-

[6]The previous FA/ITV deal assumed the screening of 18 live matches per season (see Baimbridge, Camerson, & Dawson, 1996).

ducing the act, the then Minister for Arts, Heritage, Gaeltacht, and the Islands, Síle de Valera, stressed that her intent was "to protect the major cultural and sporting events from pay-per-view." She acknowledged the concerns of event organizers regarding the impact of the legislation on their ability to negotiate television deals but countered this by noting, "it must be recognised that the citizen has rights too, and these must be safeguarded" (Mac Carthaigh, 1999, p. 53). The short (five pages long) Bill won broad support from all parties in the Dáil and passed without controversy into law in Autumn 1999.

At this point it was assumed that the next stage would be for the Minister to prepare a list of protected events as the legislation required. Yet in the nearly 3 years between the passing of the legislation and the announcement of the FAI/BSkyB deal no list was drawn up. It is impossible to say precisely why this should have been the case. One can reasonably disregard the possibility that the government did not consider it urgent, if only because the period 1999 to 2002 saw recurrent warnings of the consequences of not drawing up the list. In 1998, 2000, and again in February 2002, Irish viewers were unable to watch Ireland's rugby matches against England in the Six Nations Rugby Championship. This came about as a result of a 5-year (1998 to 2002) deal between the English Rugby Football Union and BSkyB whereby the former sold the latter exclusive live screening rights to home games played as part of the Six Nations Championship. This led one Fianna Fáil Senator to publicly criticize Síle de Valera's failure to preserve the event for Irish viewers by nominating it as a cultural event (Thornley, 2002).

When such criticism from within the government's own party was unable to prompt some action questions are raised as to what other interests might have been served by any failure to put the act into effect. Clearly these would have included the various sporting organizations that would potentially see their events included on a list—the FAI, the Irish Rugby Football Union and the Gaelic Athletic Association. Indeed the government would later explicitly state that the unwillingness of the organizations to agree to any list had been a major source of delay. However, given that the 1999 legislation's entire rationale was based around the assumption that such organizations would not voluntarily commit to selling their games to free-to-air channels, it was never clear as to why the recalcitrance of the organizations could block the drawing up of a list. In any case the professed impotence of the Irish government in dealing with the FAI was somewhat belied by the Tánaiste (Deputy Prime Minister) Mary Harney who, at the time, disclosed that substantial amounts of grant aid were being afforded to the FAI by the same government.

This then points to another possible interested party—BSkyB and its owner Rupert Murdoch. The chilling effect of Rupert Murdoch on politicians has frequently been asserted (if less frequently demonstrated). In this regard, writers on political economy of the media frequently point to UK Prime Minister Tony Blair's trip to Hayman Island, Australia in 1995 to address the "massed ranks of News Corporation executives" (Curran & Seaton, 1997, p. 300) and to meet Rupert Murdoch himself. Although the details of the private meeting were never disclosed, some press comment assumed that the two agreed a deal whereby Blair would soft-pedal restrictions on cross-media ownership (if and when he entered government) in return for guarantees relating to the editorial line of News International titles in the United Kingdom with regard to New Labour (Porter, 1995).

This is supported by Curran and Seaton's (1997) point about Murdoch's ability to "weave his way past regulatory obstacles in Australia, the United States and Britain through a combination of political horse-trading, intimidation and charm" (p. 366).

This assumption is also supported by the apparently warm relationship between Murdoch and the current Irish administration. In June 2002, a month before the signing of the FAI/BSkyB deal, Rupert Murdoch was in Ireland at the inauguration of a new News International printing plant in Kells, County Meath. He was welcomed to Ireland by the individual officially undertaking the opening—Fianna Fáil party Leader and Irish Prime Minister Bertie Ahern. In his opening address Ahern described Murdoch as "one of the world's foremost leaders in media who has been hailed as one of the most outstanding Australian figures of modern times" (Cullen, 2002, p. 5). In an intriguing response, given the temporal proximity of the television deal, Murdoch said it was hard to think of anywhere more conducive to business than the Republic (Smurfit & Murdoch, 2002).

This meeting followed 5 years of support for the Fianna Fáil party (and Taoiseach Bertie Ahern in particular) from News International titles published in Ireland. Although historically Murdoch has been at pains to stress his noninterference in the line taken by the editors of his newspapers, it nonetheless appears that the editors of the Irish editions of the *Sun*, the *News of the World*, and the *Sunday Times*, share Murdoch's high opinion of the current government. For example, one unpublished study (Daly, 2002) on the *News of the World's* political content tracked the political allegiance of those politicians invited to pen the paper's comment column, located on its editorial page. This study noted that of the 96 published comment pieces by Irish politicians between 1998 and 2000, 78 were written by Fianna Fáil TDs or senators. By contrast the next largest party—Fine Gael—was invited to write just two pieces over the same period. The study also calls attention to an interview with Bertie Ahern, then opposition leader, published by the paper on May 18, 1997, one day after the calling of a general election. Under the heading "Ahern the Man for All Ireland" the front page, and two inside pages were given over to the interview, which proclaimed that Bertie Ahern "could become a statesman to equal the legendary Eamon de Valera." However what is most striking is that the interview was conducted, not by the paper's Irish editor, but by a senior associate editor based in London, Alex Murunchak. Noting that Mr. Marunchak's by-line has not appeared in the Irish edition of the *News of the World* since the study, points to similarities with the direct intervention by Rupert Murdoch to ensure that *The Sun* endorsed Tony Blair at the outset of the 1997 British General election campaign (Daly, 2002).

Yet recent political history in Ireland has reminded politicians of how fickle press support can be, particularly when government policy clashes with business interests owned by large-scale media organizations. In this respect it is worth noting that Ireland has the greatest level of concentration of press ownership in the European Union: The newspaper market is dominated by the Independent Newspaper Group that owns titles accounting for 70% Irish Sunday sales, 100% of the evening daily market, and 49% of daily sales—including the best-selling daily in the country the *Irish Independent*.[7] On the eve of the 1997 election, this last paper

---

[7] Author's own calculations based on Audit Bureau of Circulation 2003 data. These percentages assume the Irish market is limited to domestically headquartered newspapers and thereby exclude Irish

abandoned a half-century old pro-Fine Gael (then the leading party in a coalition government) editorial line, and urged voters to support Fianna Fáil. This dramatic shift was largely driven by the coalition's failure to prosecute those involved in establishing local television distribution services in rural areas. This arguably undermined the position of Princes Holdings, a cable company then owned by Independent Newspapers. Fine Gael might have lost the election regardless of the *Independent's* editorial line but as Horgan (2001) stated, the "Payback Time [editorial] marked a watershed in Irish journalism which few would forget" (p. 171).

Nor is the influence of the News International Group in Ireland in this respect to be underestimated: the *Irish Sun* is now the second best-selling daily paper in Ireland. Meanwhile the two News International Sunday Titles—*The Sunday Times* and the *News of the World*—now account for more than 20% of all "Irish" Sunday sales (Audit Bureau of Circulation, 2003). Furthermore lest the Irish government was in any doubt as to how News International would view the furore over the deal, the Irish editions of both Sundays on the weekend following the BSkyB deal distinguished themselves by—in contrast to every other paper's editorial line—arguing that the BSkyB deal was a good thing for soccer in Ireland. A writer in *The Sunday Times'* Irish edition noted that "the FAI deserves credit for another respectable leap towards financial viability" (O'Brien, 2002, p. 15).

7. THE HALF-TIME DISCUSSION

The extended public debate following the deal quickly resolved around a number of assumptions, which were presented as commonsense. In crude terms these were that:

- The FAI were greedy
- RTÉ were cheap
- BSkyB were also greedy and culturally insensitive
- The government was inept

None of these things were straightforwardly true but they were understandable given that one had a story that could arguably only be completely told through combining the perspectives of a business correspondent, a political commentator, a sports editor, and a media journalist.

The last 15 years has clearly witnessed a transformation in the financial context in which sports organizations operate—most overtly in BSkyB's acquisition of the rights to the English Premier League.[8] In this regard Boyle et al (2002) make the point that many sports journalists are ill-equipped to report on the business of sport, lack-

editions of U.K. titles. In the case where a newspaper is only part-owned by independent newspapers I have only included sales figures equivalent to the company's shareholdings in that title—for example, 50% of the Irish Star's sales that independent newspapers co-own with the Express Group in the United Kingdom.

[8]In 1988, ITV paid the English Football League £11m for 4 years of television rights. When the rights were next negotiated in 1992, BSkyB bid—a then staggering—£191.5m for the same period of time. This rose again to £670m in 1997 (see Cave & Crandall, 2001).

ing as they do a basic knowledge of financial matters. This is reflected in the Irish context—criticisms of RTÉ's poor offer did not recognize that in economic terms subscription and free-to-air broadcasters were not operating in the same markets and that sporting rights were—by definition—worth more in a subscription market. Equally the question of whether the FAI was de facto a monopoly and could not expect to escape some regulation of its activities was not extensively discussed.

Indeed there was obviously also some confusion as to whether the story should be treated in business terms at all. Boyle et al. (2002) noted an increasing interest in the business of football across the media spectrum citing for example, *The Financial Times'* weekly sports page is written by financial journalists. However, although there is a clear logic for doing this with regard to clubs like Manchester United, clubs that are also literally PLCs turning over £100m plus (Manchester United, 2004), it is harder to determine how national sporting organizations who did not owe an allegiance to shareholders on the stock market—that is, the FAI—should be treated.

Yet if Boyle et al. (2002) concluded that many sports journalists are unable to adequately cover the business side of sports, the Irish case study suggests that the degree of subject specialization that now characterizes all journalistic work leaves many ill-equipped to deal with stories that revolve around politics, markets, and sports. In particular the Irish case study points to the difficulties thrown up by a sports story so heavily affected by the influence of domestic politics and European legislation on regulating cultural affairs.

This is evident from any examination of the government's response to the deal. Driven by a private recognition that it bore some measure of the culpability for the deal, the state sought to distance itself from the entire affair. A series of announcements from a variety of government ministers sought to ensure that the news agenda portrayed the matter as purely in the hands of the FAI, BSkyB, and RTÉ. Thus a day after the deal was announced Minister for Communications Dermot Ahern, described it as a "devastating blow" adding that it had come as a "bolt from the blue" (McNally, 2002, p. 1). Two days later again, Minister for Arts and Sports John O'Donoghue made it clear that the deal was a private one that that the government could not intervene in, describing the agreement as "done and dusted" (O'Sullivan, 2002). Having established this distance both the aforementioned Ministers and latterly the Taoiseach (who described the deal as "disenfranchising" Irish fans; Flynn, 2002c) were in a position to express, variously, outrage and disappointment at the deal.

Over the first weekend of the story this media management was partly successful: The line adopted by most journalists was that the FAI had hoodwinked RTÉ and Irish fans. As Monday dawned, however, criticism began to emerge of the government's lengthy failure to prevent the deal by listing the home internationals under the 1999 Broadcasting (Major Television Events Coverage) Act. Strikingly, the initial reference to the 1999 Act did not come from a journalist: Indeed it seems that no Irish print or broadcast journalist was aware of either the Irish legislation or the European directive on which it was based. Instead the Act was pointed to more or less simultaneously by both an opposition politician and the author of the current piece.

Given the legislation and the state's failure to use it on the agenda, the government was forced to respond. They did so in a politically novel fashion: They criticized their own legislation with a series of assertions about its limitations. These were:

- That the legislation was "legally risky".
- That it was impossible to designate entire tournaments under the legislation.
- That the consent of sporting or cultural rights holders such as the FAI was required before "their" events could be listed.
- That lists drawn up in other European countries had been rowed back on.
- That to list events under the legislation would effectively force sporting organizations to sell their television rights to RTÉ, thus undermining their bargaining position when striking deals on those rights.

However these assertions proved to be either incorrect or simply irrelevant. Items 2 and 3 were entirely inaccurate, and were apparently made without any reference to the relevant legislation. The point about the inability to designate entire tournaments stood out in this regard. There was nothing in either the Irish legislation, EU directives, or actual practice in other countries, which could be remotely construed in such a manner. The first assertion describing the legislation as legally risky was clearly a red herring—no legislative proposition is ever entirely without such ambiguity. Meanwhile the final assertion—although true—missed the core (indeed almost only) intent of that Broadcasting Act as explicitly stated by the minister who introduced the legislation in 1999: to ensure that the financial interests of cultural organizations were not placed above the rights of citizens to participate in cultural events.

Once again most journalists simply accepted these assertions regarding both the Irish and EU legislation as fact. There is a substantial literature on the extent to which journalists rely (and given deadline pressures arguably must rely) on official sources. However the Irish case study suggests that the state actively exploits this fact, making misleading assertions on the assumption that no journalist will have the time to check them out, particularly if the story is one that involves dealing with several specializations simultaneously.

However, this reliance on authoritative sources can work in more than one way. Six days into the story the current author wrote an opinion piece for *The Irish Times* that tentatively suggested that although the government had apparently missed the boat in terms of listing the Irish games that there was nothing in the domestic legislation specifically prohibiting the designation of games already subject to a deal between a sporting organization and a subscription broadcaster (Flynn, 2002b). Two days later *The Irish Times* and RTÉ carried interviews with a senior European Commission official attached to Viviane Reding's Education and Culture Directorate. The official, Christophe Forax, asserted that a list once published applied "to all events listed—even contracts signed beforehand" (Hennessy, 2002, p. 1).

In passing one should note that, given the prior existence of the FAI/BSkyB deal, such designation would indeed have been "legally risky". Although Article 3a of TWF allows governments to draw up lists, it notes that they should do so "in a clear and transparent manner *in due and effective time*" (italics added). Such retrospective designation also appeared to clash with the Irish constitution's stout defense of property rights. However the straightforward acceptance of Christophe Forax's comments by the Irish media pointed to another failure to actually read the European legislation. As a consequence, however, the government suddenly found it politically necessary to reconsider their position. Faced with an issue that was not going away the government apparently decided that their best bet was to adopt the role of the white knight belatedly rescuing Irish fans. Hence on July 13, days after his Minister for Sports' "done and dusted" comment, the Taoiseach presented the government's new line:

> We are the sovereign government and there is legislation and an EU directive and it is quite clear. We cannot sit around and disenfranchise a large part of the population from seeing matches they are allowed to see. (Flynn, 2002c, p. 144)

Yet the manner in which the government actually sought to turn the deal around suggests that the state was well aware of the very real legal risk in attempting to effectively render the FAI/BSkyB deal null and void.

8. EXTRA TIME

The government followed the announcement of their determination to wrest back the games for Irish citizens by calling a series of private meetings with the country's major sporting organizations informing them of their intent to designate some of their games. Veiled references were made to the Attorney General's view that the state had some "legal options" regarding the deal. Predictably, the organizations objected to any suggestion that deals might be undone by lists that had not been in existence at the time when the original deals were signed. However, what made these meetings with the sporting organizations unusual is that both Article 3a of Television Without Frontiers and the 1999 Irish legislation refer to the power of member states/governments to dictate to broadcasting organizations with regard to designated events:

> Each Member State may take measures in accordance with Community law to ensure that *broadcasters* under its jurisdiction do not broadcast on an exclusive basis events which are regarded by that Member State as being of major importance for society. (European Council, 1997, p. 66)

> Where a *broadcaster* under the jurisdiction of the state who is not a qualifying broadcaster acquires rights to broadcast a designated event, that broadcaster shall not broadcast the event unless the event has been made available to a qualifying [i.e. free-to-air]. (Government of Ireland, 1999, p. 6)

If there was no reference to event organizers in this or any other article in the 1999 Irish legislation, what then was the point of these meetings?

The question was only finally resolved by the second major innovation introduced by the Irish state with regard to the workings of TWF (the first being the notion that it permitted retrospective going forward designation of events). In

October 2002, after months of consultation, the Irish government published a list of designated events, which included the Republic's home and away soccer European Championships and World Cup qualifying matches, the All-Ireland hurling and football finals, the Irish Derby, and the Irish Grand National. However they also announced that they would need to amend the 1999 legislation to ensure that the games covered by the FAI/BSkyB deal would be made available on a free-to-air basis.

That amendment appeared in April 2003 via the Broadcasting (Major Events Television Coverage Amendment) Act. The second article of the new legislation covered the question of retrospection:

> This Act applies to a designated event, which is designated before or after the passing of this Act ... *whether or not any agreement or arrangement has been entered into* between the event organizer and a broadcaster in respect of the acquisition by the broadcaster of the rights to the event. (Italics added; Government of Ireland, 2003, p. 4)

However the most significant change came with Article 4 which now read:

> Where an event has been designated ... and if ... the *event organiser* has not made an agreement or arrangement with a qualifying broadcaster to enable it to provide coverage on free television services with the state ... a qualifying broadcaster may apply to the High Court ... for an order directing the *event organiser* to give rights to the qualifying broadcaster to provide such coverage. (Italics added; Government of Ireland, 2003, p. 5)

In short the amendment shifted the focus of the exercise of the state's authority from the broadcasting organizations to the events organizers. In effect while the 1999 legislation would have brought the state into direct confrontation with BSkyB, the amendment set up a showdown with the Football Association of Ireland. In so doing the 2003 Amendment appears to exceed the authority granted to member states by Article 3a, which does not appear to extend beyond authority over broadcasters.

The significance of this shift was entirely missed by the Irish media—not one report noted the changes in focus between the 1999 and 2003 legislation. However, it is also not entirely clear as to whether the European Commission itself was aware of this shift when it gave approval to the Irish list (as it would do early in 2003). In Ireland's notification to the Commission published in the EU's Official Journal in April 2003, no reference is made to the Amendment Act. Instead the Commission's approval was based on a list drawn up under the 1999 legislation (CEC, 2003).

Yet it was the 2003 legislation that was used to enforce the current status quo. As of 2004 the FAI/BSkyB deal has been renegotiated so that Ireland's qualifying games at least are screened on both RTÉ (which eventually paid out €2.3m for the rights (Malone, 2003) and BSkyB Sports. Live rights to Ireland's other home games are still the exclusive preserve of BSkyB although Irish fans can view deferred coverage on TV3.

However, if Christophe Forax's July 2002 assertion that Ireland's existing legislation was sufficient to permit retrospection we need to ask why this shift took place? One possible explanation is that it offered the Irish state a means of retrieving (some of) the Irish games at an acceptable political cost. That the current Irish administration might be reluctant to take on an organ of News International has already been noted. By contrast offending the FAI—a body already the villain of the piece in the public eye—was tolerable (especially given that parallel negotiations between the Irish government and the FAI relating to the building of a national stadium offered the government a certain degree of leverage in negotiating the retrospective designation of events. Although the BSkyB's €7.5m offer to the FAI represented a significant sum, it was dwarfed by the possible state investment into a national stadium).

Yet if political self-interest offers a partial explanation for the introduction of the 2003 Amendment, the state itself would offer a more explicit statement of its thinking in a document submitted to the European Commission in Summer 2003. Following the publication by the European Commission of the Fourth Review of the Television Without Frontiers directive (CEC, 2003a) in January 2003, the Commission invited written contributions from interested parties on the conclusions of the Review.

The submission from the Irish State is one of the longest. The major thrust of the submission argues for a radical reinterpretation of Article 2 of the TWF directive, which denies Member States the right to restrict retransmission on their territory of television, broadcasts from other Member States. The Irish submission notes that in the era of cross-border broadcasting the:

> ability of some Member States to have meaningful regulation at a national level will be eroded unless the Directive is revised to provide that services directed primarily at a given Member State are subject to the regulation of that Member State. Ireland suggests that this should be achieved by determining a competent jurisdiction for each individual broadcasting service, instead of focusing on broadcasters. (Department of Communications, Marine & Natural Resources [DCMNR], 2003, p. 2)

In short, the Irish submission argued that if a U.K. registered company established a station explicitly targeting the Irish market then it should be regulated by the Irish state not a British regulator. As such the Irish submission suggests a reversal of one of the original directive's most fundamental tenets – that individual states should not have the power to regulate broadcasting services that were considered acceptable under another state's broadcasting regime.

This point forms the context for the submission's comments on Article 3a, which also offer some clarification for the content of the 2003 legislation. (These are lengthy but deserve quoting in full):

> a UK broadcaster only requires the consent of the Independent Television Commission to purchase exclusive rights to an event after an event has been formally designated in another Member State. Thus according to the ITC code, it was possible for BSkyB to enter into a contract for exclusive live rights for Ireland's qualifying games in the Football World Cup and European Championship games without the consent of the ITC. However the subsequent designation of the events by Ireland did not, it would appear, require BSkyB to seek the approval of the ITC as the contract had already been entered into. There was therefore no retrospective effect.
>
> In order to ensure that these games were made available to Irish free-to-air broadcasters, Ireland enacted new legislative provision which imposed obligations on sports

> event organizers to make broadcasting right available to qualifying broadcasters subject to the payment of reasonable market rates. (DCMNR, 2003, p. 5)

The reference to the ITC is at the crux of the matter. As a U.K. registered broadcaster BSkyB was subject to the U.K.'s Independent Television Commission. Clearly the Irish government had little faith in the ITC's ability/willingness to pursue BSkyB. However, this seems overly pessimistic given the existence of a prior case where the ITC had gone to bat on behalf of another member state. In June 2000 TVDanmark, a Danish subscription broadcaster outbid two Danish public broadcasters to secure exclusive rights to five Danish World Cup qualifying matches. These were designated events under Danish legislation but the Danish authorities were unable to force TVDanmark to make the games available to either of Denmark's public service channels because TVDanmark was registered as a British company. As such it came within the regulatory ambit of the Independent Television Commission in the United Kingdom who went to House of Lords, the highest legal authority in the United Kingdom in an effort to force TVDanmark to offer the games on a nonexclusive basis to Denmark's public service broadcasters. TVDanmark, however, argued that it had acquired the rights to the Danish games in accordance with the ITC's own rules, a point conceded by the ITC. However, the ITC had also consulted with the Danish Ministry of Culture, which noted that under the Danish system for protecting listed events TVDanmark's acquisition of the rights would not have been legal. Consequently, the ITC decided to pursue TVDanmark under the Danish code: in effect a regulatory authority from one European country applied the—slightly different—rules of another regulatory authority in its own jurisdiction, a decision that House of Lords implicitly endorsed. The Lords ruling stated that the intention of Article 3a was "perfectly clear", namely:

> To prevent the exercise by broadcasters of exclusive rights in such a way that a substantial proportion of the public in another member state was deprived of the possibility of following a designated event. (House of Lords, 2001, p.1)

Given prior evidence that the ITC was willing to go to bat on behalf of designated events in other member states, why was the Irish state unwilling to pursue that route? This reluctance seems likely to have been prompted by concern over the question of retrospective designation. In the TVDanmark case, it appears that the ITC, while acknowledging the differences in the manner in which the United Kingdom and Denmark had implemented Article 3a, felt that that there was nothing in the Danish legislation that actually contravened Article 3a. However, the same may not have been true of the Irish attempt to effect a retrospective "going forward" listing. As noted earlier, Article 3a stresses the need to draw up such a list in a timely fashion. This is reflected in the British Broadcasting Act 1996, Section 101 of which was amended to effect Section 3a of TWF. Subsection 4 of Section 101 noted that designations of events would "not have effect where the television program providing the first service is exercising rights acquired before the commencement of this section." That is retrospective designation of events was not permitted.

Hence the Irish government passed a piece of legislation the preamble of which states that it will give "further effect to article 3a" but that appears to contravene aspects of precisely that article, by permitting retrospection and focusing on the obligations of event organizers rather than broadcasters.

9. THE FINAL WHISTLE

The major question raised in the Irish public sphere by the BSkyB deal focused on the how, and where, one draws the line between sport as cultural event and sport as business. The role of the national side as an expression of national identity creates a strong attachment to that team from the national audience. This has implications for public service broadcasters. As one national newspaper commentator recognised with regard to the BSkyB deal: "What is particularly corrosive about this development for RTÉ is that it undermines its whole identity as a national broadcaster" (Oliver, 2002). If public service broadcasting is meant to mean anything, it is meant to mean broadcasters reflecting and covering truly national events that bind the country together. The loss of such important cultural or sporting events makes RTÉ's attempts to remain a truly national broadcaster extremely difficult.

However, the almost religious fervor excited by the Irish team among its home fans translates in commercial terms into content, that can draw in audiences in a more or less guaranteed fashion. The scale of the vitriol hurled at the FAI seems in large part to have been motivated by the fact that the FAI's actions reminded fans of the painful fact that commerce plays a considerable role in modern sport. This may be overt in which the team in question is a PLC—the question is whether this should be case in dealing with national teams.

This confusion was reflected in press coverage at the time of the deal and as noted earlier in the failure to simultaneously examine the story from sporting, political, and financial perspectives. However such failings were not simply the result of difficulties over journalistic demarcation. Although most of Ireland's major daily and Sunday papers criticized the BSkyB deal, most did so by treating BSkyB as a standalone company. The significance of cross-media ownership in the BSkyB deal was ignored by large elements of the Irish press: titles owned by the market-dominating Independent Newspapers Group seemed particularly reluctant to discuss the link between concentration of media ownership within and across media and political influence. Ironically, the FAI/BSkyB deal was a story that could only be completely understood by drawing attention to a subject—concentration of media ownership—that a major element of the Irish public sphere has an active interest in avoiding. The dominance enjoyed by Independent Newspapers did not emerge from a vacuum, however one can argue that much of the weakness of the Irish public sphere derives from a consistent failure on the part of Irish governments to take regulation of media ownership seriously.

However the story also points up a weakness in another element of the Irish public sphere, RTÉ, one for which the state is more directly culpable. Following a major showdown between the state and RTE in 1972, which ended in the sudden removal of the RTE authority, the Irish public service broadcaster has generally been quiescent in its dealings with the government of the day. The situation whereby RTÉ's license fee was effectively frozen from 1986 to 1996 is thus understood "locally" as an attempt by the state to keep the broadcaster on a tight leash (Corcoran, 2004, pp. 98–103). The idea of a direct broadcaster–state confrontation along the lines of the BBC's recent clash with the Blair administration over the death of weapons expert David Kelley, for example, is almost inconceivable in an Irish context. In this respect the timing of the BSkyB deal was particularly awkward for RTÉ, given its coinciding with delicate negotiations between the government and broadcaster over precisely the question of a license fee increase. Such conditions were not conducive to the production of programs discussing either the link between RTÉ's financial position and their low offer to the FAI or broader issues such as the link between the government and News International. On the surface RTÉ's situation with respect to future license fee increase has subsequently been placed on a more secure footing: in the aftermath of the FAI/BSkyB deal RTÉ did in fact secure a substantial license fee increase. What is more they succeeded in convincing the government to index future increases to inflation. However such index-linked increases are not automatic but are subject to the government's assessment that RTÉ is meeting certain restructuring/public service requirements. In effect, future increases remain in the gift of whatever administration happens to be in power.[9] The implications for the objectivity of the Irish public service broadcaster in covering the state's activities are worrying.

Ultimately, however, the most important issue may be that thrown up by the discussion in the latter part of this article, namely the extent to which TV Without Frontiers fails to address the issue of transnational services dedicated to a foreign state, a situation clearly not contemplated in the framing of the directive in 1989 or 1997. It has subsequently emerged that the Irish 2003 submission on TWF was simply the first salvo in a campaign to place the regulatory challenge of transnational broadcasters at the center of the broadcasting policy agenda during Ireland's presidency of the EU in the first 6 months of 2004. Outlining his policy priorities for the presidency, the Irish Minister for Communications noted that:

> We need a level playing field in the area of broadcasting. It is simply not on that satellite broadcasters are not regulated in the markets into which they broadcast. We need to be able to apply national measures to satellite broadcasters. (Government Information Service [GIS], 2004a)

Specifically the Irish government has sought to table an amendment to TV Without Frontiers, which would allow governments to regulate broadcasters who provide television services specifically targeted at viewers in another country. However, at an informal EU Ministerial Conference on Broadcasting held in Dublin in March, the Irish proposal met resistance from both the European Commissioner Viviane Reding and indeed several other member states (GIS, 2004b). Although the Irish government's press release after the meeting suggested that the question was still open, Viviane Reding's comments suggest otherwise:

---

[9]Actual practice bears this concern out. Although RTE was technically entitled to a 2.2% increase in its license fee at the close of 2003, the government capped this at 1.5% citing concerns over RTÉ's failure to sufficiently reduce staff numbers (see Oliver, 2003).

> The "originating country" was responsible for regulation and licensing. "That is the general rule of the common market. It is not possible to change that," she said. (Oliver, 2004, p. 11)

Given this outcome the essentially stopgap measure that was the 2003 Irish Broadcasting legislation looks set to remain in place creating an unusual (if not unique) negotiating environment for Irish sporting organizations. As both the FAI and Irish Rugby Football Union pointed out in their own submissions to the European Commission[10] with regard to the retrospective going forward of Section II of the Broadcasting Bill 2003:

> The way Section II is constructed means that there is no certainty from the point of view of anybody dealing with the Association in relation to contractual arrangements we conclude in the future for the broadcasting of our matches. At any stage in the future the Minister can create a new classification for matches other than qualification matches that we organize which would then retrospectively impact on any contractual arrangements we had already made in respect of those matches. From a commercial point of view this creates an unacceptable level of certainty for any broadcasters we may be dealing with. (Football Association of Ireland, 2003; Irish Rugby Football Union, 2003)

It is hard to imagine that the original framers of Article 3a ever envisaged such an application of the article.

REFERENCES

Audit Bureau of Circulation. (2003). *Circulation review July – December 2003 v 2002.* Dublin, Ireland: ABC.

Baimbridge, M., Camerson, S., & Dawson, P., (1996). Satellite television and the demand for football: A whole new ball game? *Journal of Scottish Political Economy, 43,* 317–333.

Boyle, R., Dinan, W., & Morrow, S. (2002). Doing the business: Newspaper reporting of the business of football. *Journalism, 3,* 161–181.

Byrne, P. (1996, October 24). RTE to cover all Ireland's home ties. *The Irish Times.*

Cave, M., & Crandall, R. W. (2001). Sports rights and the broadcast industry. *The Economic Journal 111,* F4–F26.

Commission for Communications Regulation. (2003). *Irish communications market: Quarterly key data December 2003.* Dublin, Ireland: Comreg.

Cullen, P. (2002, July 12). Shark who likes football gets Bertie's good press. *The Irish Times,* p. 5.

Curran, J., & Seaton, J. (1997). *Power without responsibility* (5th ed.). London: Routledge.

Daly, P. (2002). *The political coverage of the Irish edition of the news of the world 1997 – 2000.* Unpublished manuscript, Dublin City University, Ireland.

Department of Communications, Marine & Natural Resources. (2003). *Ireland's submission to the European Commission in relation to the review of the Television Without Frontiers Directive.* Retrieved April 5, 2004, from http://europa.eu.int/comm/avpolicy/regul/review-twf2003/wc_ireland.pdf

Dooley, C. (2002, July 26). FAI firefighter gets plenty of chances to put out flames. *The Irish Times,* p. 60.

Commission of the European Communities (CEC). (2002). Publication in accordance with Article 3(a)(2) of Council Directive 89/552/EEC (1) on the coordination of certain provisions laid down by law, regulation or administrative action in Member States concerning the pursuit of television broadcasting

---

[10]Interestingly Satellite and Cable Broadcasters' Group (which included British Sky Broadcasting) also made a submission to the Commission but made no reference to Article 3a or the Irish events.

activities, as amended by Directive 97/36/EC of the European Parliament and of the Council. *Official Journal C 016*, pp. 8–10.

Commission of the European Communities (CEC). (2003a). *Fourth report from the commission to the council, the European Parliament, the European economic and social committee and the committee of the regions on the application of directive 89/552/EEC "Television without Frontiers"* COM (2002) 778 Final. Brussels, Belgium: Author.

Commission of the European Communities (CEC). (2003b). Publication in accordance with Article 3a(2) of Council Directive 89/552/EEC on the coordination of certain provisions laid down by law, regulation or administrative action in member states concerning the pursuit of television broadcasting activities, as amended by Directive 97/36/EC of the European Parliament and of the Council. *Official Journal C 100*, 12–16.

Corcoran, F. (2004). *RTE and the globalisation of Irish television.* Bristol, England: Intellect.

European Council. (1989). Council directive 89/552/EEC of 3 October 1989 on the coordination of certain provisions laid down by law, regulation or administrative action in member states concerning the pursuit of television broadcasting activities. *Official Journal L 298*, 23–30.

European Council. (1997). Council directive 97/36/EC of the European Parliament and of the council of 30 June 1997 amending council directive 89/552/EEC on the coordination of certain provisions laid down by law, regulation or administrative action in member states concerning the pursuit of television broadcasting activities. *Official Journal L 202*, 60–70.

Flynn, R. (2002a). Broadcasting and the Celtic tiger: From promise to practice. In P. Kirby, L. Gibbons, & M. Cronin (Eds.), *Reinventing Ireland.* (pp. 160–176). London: Pluto.

Flynn, R. (2002b, July 9). Government must not be afraid to tackle vested interests. *The Irish Times*, p. 12.

Flynn, R. (2002c, August), Kicking to touch? *Magill*, p. 25.

Football Association of Ireland. (2003). Re: Review of the television without frontiers directive. Retrieved April 5, 2004, from http://europa.eu.int/comm/avpolicy/regul/review-twf2003/wc_fai.pdf

Foley, M. (1993, January 14). Cablelink seeks ads on satellites. *The Irish Times.* p. 5.

Government Information Service (GIS). (2004a, January 14). Dermot Ahern outlines presidency priorities. Dublin, Ireland: Author.

Government Information Service. (2004b). *Informal ministerial conference on broadcasting, March 2–3: Outcome of proceedings, prepared by Irish Presidency.* Dublin, Ireland: Author.

Government of Ireland. Broadcasting (Major Events Television Coverage) Act 1999. (1999).

Government of Ireland. Broadcasting (Major Events Television Coverage) (Amendment) Act 2003. (2003).

Grainne Seoige to present new Sky news service for Ireland (2004). *In Production, 7*, p. 1.

Hennessy, M. (2002, July 11). EU official says Sky deal can be bypassed. *The Irish Times*, p. 1.

Horgan, J. (2001). *Irish media, a critical history since 1922.* London: Routledge.

House of Lords. (2001, July 25). Lord Hoffman, Regina vs. Independent Television Commission, Ex Parte TVDanmark 1 Ltd. House of Lords.

Humphries, T. (2003). *Laptop dancing and the nanny-goat mambo: A sportswriter's year.* Dublin, Ireland: Townhouse.

Irish Rugby Football Union. (2003). *Re: Review of the television without frontiers directive.* Retrieved April 5, 2004, from http://europa.eu.int/comm/avpolicy/regul/review-twf2003/wc_irfu.pdf

Mac Carthaigh, S. (1999, June 11). Minister to tackle threat to cultural identity as digital age takes hold. *The Irish Times*, p. 53.

Malone, E. (2003, June 17). Broadcasting deals do little for home game. *The Irish Times*, p. 25.

Manchester United. (2004) *Investor Relations Financial Date 5 Year Summary.* Retrieved April 5, 2004, from http://ir.manutd.com/manutd/findata/trends/

McKie, D. (1998). Swingers, clingers, waverers, quaverers: The tabloid press in the 1997 general election. In I. Crewe, B. Gosschalk, & J. Bartle (Eds.), *Why labour won the general election of 1997.* London: Cass.

McNally, F. (2002, July 6). FAI defends E7m Sky deal for soccer ties. *The Irish Times*, p. 1.

Ministry of Culture–Denmark. (2003). *Consultation on the television without frontiers directive.* Retrieved April 5, 2004, from http://europa.eu.int/comm/avpolicy/regul/review-twf2003/wc_denmark%20ministry%20of%20culture_en.pdf

A. C. Neilsen. (1997). *September 1997 RTE TV review.* Retrieved April 5, 2004, from http://www.medialive.ie

A. C. Neilsen. (2004). *February 2004 RTE TV Review.* Retrieved April 5, 2004, from http://www.medialive.ie

O'Brien J. (2002, July 7) At last, a good bit of business by FAI. *The Sunday Times (Irish ed.)*, p. 15.

Oliver, E. (2002, July 6). "Green Army" being asked to tune in, turn on and pay up. *The Irish Times*, p. 5.
Oliver, E. (2003, December 17). RTE chief criticises higher fee refusal. *The Irish Times*, p. 9.
Oliver, E. (2004, March 3). Battle for satellite TV control lost. *The Irish Times*, p. 11.
O'Sullivan, J. (2002, July 10). RTE bid not in ballpark say FAI. *The Irish Times*, p. 18.
Porter, H. (1995, July 18). Zealous moderate. *The Guardian*, p. T2.
Radio Telifís Éireann Authority. (2002). *Forty-second report and statement of accounts for the 12 months ended 31 December 2002,presented to the minister for communications, marine and natural resources pursuant to sections 25 and 26 of the Broadcasting Authority Act, 1960*. Dublin, Ireland: Author.
Smurfit, Murdoch printing plant opens. (2002, June 8). *The Irish Times*, p. 19.
Thornley, G. (1994, September 17). RTE and FAI sign four year contract. *The Irish Times*, p. 19.
Thornley, G. (2002, February 12). The long-term costs of pay-per-view. *The Irish Times*, p. 21.

For Product Safety Concerns and Information please contact our EU representative GPSR@taylorandfrancis.com Taylor & Francis Verlag GmbH, Kaufingerstraße 24, 80331 München, Germany